Palgrave Studies in Logistics and Supply Chain Management

Series Editors
Martin Christopher
School of Management
Cranfield University
Cranfield, Beds., UK

Emel Aktas ⓘ
School of Management, Building 32
Cranfield University
Cranfield, Bedfordshire, UK

Logistics and supply chain management has always been critical to business success and to the efficient running of economies and the well-being of the societies they serve. In recent years, however, the importance of these activities has been highlighted by geopolitical challenges and uncertainty that have threatened global supply chains.

The series will provide an opportunity to explore in-depth and with rigour topics such as digital transformation, supply chain risk and resilience, sustainability and the circular economy, and the future of global supply chain design and management. Entries in the series will be international in outlook, providing fresh and innovative approaches to pressing issues in logistics and supply chain management.

Alp Yildirim • Hendrik Reefke
Emel Aktas

Mobile Robot Automation in Warehouses

A Framework for Decision Making and Integration

Alp Yildirim
Cranfield University
Cranfield, UK

Hendrik Reefke
Cranfield University
Cranfield, UK

Emel Aktas ⓘ
Cranfield University
Cranfield, UK

Palgrave Studies in Logistics and Supply Chain Management
ISBN 978-3-031-12306-1   ISBN 978-3-031-12307-8   (eBook)
https://doi.org/10.1007/978-3-031-12307-8

This Palgrave Macmillan imprint is published by the registered company Springer Nature Switzerland AG.
The registered company address is: Gewerbestrasse 11, 6330 Cham, Switzerland

Alp dedicates this book to his beloved family, who were always there for him. Hendrik dedicates this book to all his past and future students, who continue to be a source of inspiration and will hopefully benefit from this book. Emel would like to dedicate this book to her parents, Halime and Akif, who have always encouraged her to pursue her ambitions.

PREFACE

This book illustrates the applications of mobile robot systems in warehouse operations with real-life examples and discusses how they can be implemented by developing a conceptual managerial decision framework. Warehouses and distribution centres were considered the least desirable part of supply chains since most of their activities are non-value-added. However, decision-makers realised the growing costs and operational burden warehouses bring, and from that point on, warehouse automation has started receiving increasing attention. Mobile robots that carry products, boxes, cases, pallets, and even shelves were developed by hundreds of companies after the 2010s. They are used in many warehouses, and many more are evaluating alternatives to implement them in different warehouse operations.

We see books that mention warehousing and, to some extent, warehouse automation. However, to our knowledge, none of them focuses on mobile robots. Therefore, we aim to shed light on mobile robot systems by demonstrating and evaluating their characteristics with realistic scenarios and decision-making methods such as the 'Full Consistency Method'.

We prepared this book for both academia and practice. For academia, we explain the systematic literature review methodology we followed to produce the state-of-the-art mobile robot automation for reproducibility. We also provided a research agenda which mentions focus areas such as facility layout planning and robot fleet sizing that require further attention to aid the efficiency of practical applications of such solutions.

For practice, we present ways to assess mobile robot systems using multi-criteria analysis, which can be helpful to warehouse managers in identifying, evaluating, and choosing candidate systems. Moreover, the managerial decision framework covering decisions at strategic, tactical, and operational levels in detail helps decision-makers to implement a mobile robot solution step-by-step. We put particular emphasis on change management and the operational management of mobile robots through algorithms.

Cranfield, UK   Alp Yildirim
Cranfield, UK   Hendrik Reefke
Cranfield, UK   Emel Aktas

CONTENTS

About the Authors

Alp Yildirim After graduating from industrial engineering (2011) and doing an MBA (2014), Alp co-founded an innovative manufacturing company named Sanayi Sepeti LTD, which was backed by two angel investors. This experience helped Alp to become a CEO Office member in OPLOG Operational Logistics Inc., the fastest growing technology company in Turkey in 2017. Having worked there for two years, Alp decided to pursue an academic career in supply chain management and got accepted to the Leadership and Management PhD programme in Cranfield University with full bursary and scholarship.Alp's current research topic is mobile robot automation and throughput optimisation in warehouses. He tries to develop a framework for the decision-makers to aid them choose and fully integrate the chosen mobile robot system to their warehouses. His work was recently awarded the Best Research Proposal of the Advances in Production Management System (APMS) Conference 2021.

Hendrik Reefke has held academic positions in the UK, Germany, and New Zealand. Prior to this he worked in the automotive sector with roles in engineering and procurement and in project management. Hendrik is known for his capability in warehousing, a specialised area characterised by complexities regarding design, equipment, operations, and managerial implications.He holds an award-winning PhD from the University of Auckland as well as an MCom and BCom (Honours) in operations and supply chain management.Hendrik embraces a variety of methodological research approaches including simulation, process design, modelling, surveys, group decision techniques, case studies, and conceptual theory building. His work has been published in academic journals, book chapters, and conferences.Hendrik is Senior Lecturer in Supply Chain Management and Course Director of the full-time Logistics, Procurement and Supply Chain Management MSc course at Cranfield School of Management. He is an active researcher, focusing primarily on sustainable supply chain management, service supply chains, as well as performance measurement and reporting.He leads programmes in the areas of humanitarian logistics, sustainability, and operations management and has conducted various consulting and research projects, including Sustainable Transport in the UK, Health Supply Chain transformations in developing countries, Supply Chain Trends, and the effects of Brexit on Supply Chain Locations.

Emel Aktas has BSc, MSc, and PhD degrees in industrial engineering from Istanbul Technical University, Turkey. She began her career at Istanbul Technical University as a research and teaching assistant. She worked as a visiting researcher at Wayne State University, USA, and as a lecturer at Dogus University, Istanbul, Turkey, during her PhD studies.She took part as a researcher in public and private funded projects on location selection, shift scheduling, and transportation master plan strategy. Her refereed articles have appeared in a variety of journals including *European Journal of Operational Research, Interfaces, Supply Chain Management: An International Journal, Socio-Economic Planning Sciences,* and *Transportation Research Part A: Policy and Practice.* Before joining Cranfield, Emel was the course director of MSc Global Supply Chain Management programme at Brunel University Business School.Emel is working on minimising carbon emissions in urban logistics with focus on the trade-off between service levels and fuel consumption. A parallel line of research is about sequence-dependent flowshop scheduling where decisions about machine processing speed affect both service levels and energy consumption. Her research interests can be summarised as logistics and transportation, supply chain decisions, mathematical modelling, and optimisation.

LIST OF FIGURES

LIST OF TABLES

CHAPTER 1

Introduction

Fulfilling customers' orders swiftly and efficiently whilst managing the increasing complexity and variability in these orders is an ongoing challenge in warehouses. Retail sectors such as consumer electronics, clothing and footwear, grocery, or pharmaceuticals are characterised by large assortments of small products. The increasing tendency of customers to buy online leads to thousands of daily orders which, coupled with thousands of different products, can make the warehouse environment chaotic and complicate operations such as receiving, picking, sorting, and packing. Moreover, the order fulfilment systems are not always flexible and scalable enough to respond to fluctuating customer demand. During special retail days such as Black Friday, retailers' daily revenue could increase by up to fivefold, which convolutes the operations even more (*Adobe Holiday Shopping Trends* 2019). In addition, many B2B (business-to-business) and B2C (business-to-consumer) customers expect their orders to be fulfilled in minimal lead time, necessitating a flexible and scalable warehouse operation.

Due to intense competition and the complexity of orders, companies are re-engineering their warehouse operations towards automated and optimised fulfilment systems. The main driver for the shift towards automation is the need for increased fulfilment speed, which necessitates more efficient and faster operations compared to manually operated warehouses

(Wurman et al. 2008). Automation could be applied to every warehouse operation once these operations are identified and defined.

Bartholdi and Hackman (2019) define warehousing processes based on the direction of physical flows: inbound and outbound. Inbound operations consist mainly of receiving, put-away, and storage, whereas outbound processes include picking, packing, and shipping. Frazelle (2016) includes sortation as a main outbound activity of warehouses, whereas Rushton et al. (2014) consider replenishment as well (Fig. 1.1). Order picking, which is the process of moving items from storage to shipping to meet a specific demand, is generally the main operation that influences the design of a warehouse (Frazelle 2016). The order picking phase may account for 50–55% of the total operating costs of manually operated warehouses (Bartholdi and Hackman 2019; De Koster et al. 2007; Rushton et al. 2014). This high percentage is primarily due to the amount of time lost by pickers walking from shelf to shelf, looking for the necessary items to pick (Boysen et al. 2019). This manual travelling activity takes around 55% of the picker's total time (Frazelle 2016). Mobile robot automation systems were developed from 1950 onwards to eliminate the unproductive travelling time of human workers. *Cambridge Dictionary*

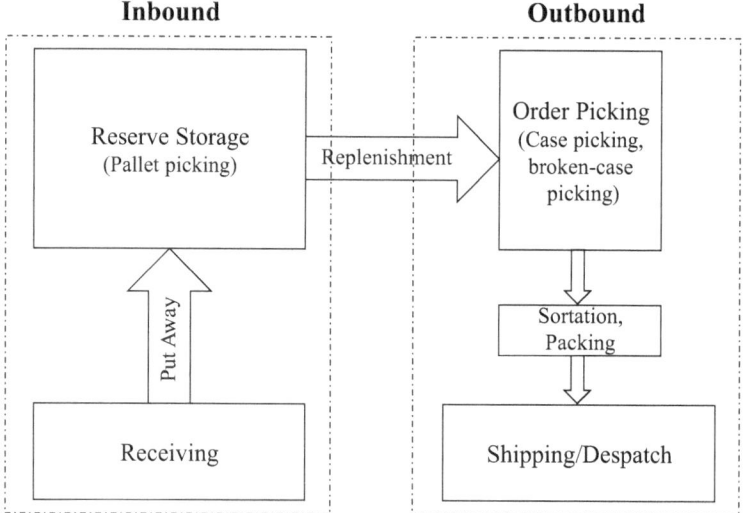

Fig. 1.1   Steps of warehousing. (Adapted from Frazelle (2016))

(n.d.) defines a robot as: 'a machine controlled by a computer that is used to perform jobs automatically'. Similarly, an unpersonned mobile robot is a mobile version of a robot. The first unpersonned mobile robot automation is the well-known automated guided vehicles (AGVs). Since 2006, autonomous mobile robots (AMRs) have become an alternative to AGVs, offering more flexible and scalable solutions (Wurman et al. 2008). These robots have onboard intelligence that helps them decide whether to slow down or stop and where to go next (Kattepur et al. 2018). They can map the environment, plan their paths, dynamically respond to their surroundings, and bypass an obstacle without being remotely controlled (Horňáková et al. 2019). These functions make AMRs more flexible and suitable than AGVs for chaotic and complicated warehouse settings such as e-commerce.

Amazon, one of the most successful e-commerce companies, automated fulfilling e-commerce orders in their warehouses with AMRs by acquiring Kiva Systems in 2012. Alibaba, Ocado, and JD.com are among the first companies that followed the warehouse automation move. With this shift in order picking operations, Amazon has cut its operational warehouse costs by 20% and saved $22 million per annum in picking costs for each of its fulfilment centres (Bogue 2016). It achieved this saving through AMRs that bring shelves full of various items to the pickers while pickers stay in the packing station to prepare orders from items brought to them in a goods-to-picker (G2P) context (Wurman et al. 2008).

G2P is a material handling context where materials are brought to the human picker by an automated system (Fig. 1.2). Pickers stay stationary during picking and other related tasks, while retrieved units are

	Picker-to-goods (P2G)	Goods-to-picker (G2P)	Robot-to-goods (R2G)
Who goes to the picking area?	Human or Human & mobile robots	Mobile robots	Mobile robots
Who picks the products?	Human	Human	Mobile robots
Example	Human collaborative robots (cobots)	Shelf-carrying robots (barcode guided)	Mobile picking robots

Level of automation ⟶

Fig. 1.2 Types of material handling contexts. (Source: Authors)

automatically restored to their inventory positions (Azadeh et al. 2019; Huang et al. 2015). The second type is the picker-to-goods (P2G) method. In this method, human pickers travel to the materials generally with a material handling unit and do the picking (De Koster 2018). Human collaborative robot (co-bot) solutions are examples of this method. Recently, a third type of material handling context has become known as robot-to-goods (R2G). Mobile picking robots are examples of this method, in which robots travel to the shelves alone and pick materials on their own (Huang et al. 2015).

Owing to the automation successes of early adopters, mobile robot adoption in warehouses had a spectacular boost. Grey Orange, Hikvision, Geek+, IAM Robotics, 6 River Systems, Magazino, and many other start-up companies are developing autonomous mobile robots for warehouses. Industry forecasts estimate that more than 50,000 warehouses (up from 4000 warehouses in 2018) will be using over 4 million mobile robots globally by 2025 (*50,000 Warehouses to Use Robots by 2025 as Barriers to Entry Fall and AI Innovation Accelerates* 2019).

Based on these developments and forecasts, Amazon's shelf-carrying robots are only one example of the many potential applications of mobile robot automation. Further, order picking is not the only warehouse operation that warrants such automation, and other warehouse areas are being targeted. To guide scholars and practitioners, mobile robot automation solutions and their usage areas in warehouses should be identified and classified. Moreover, such highly productive systems require a rationale for system selection to justify the capital investment associated with automation (Azadeh et al. 2019; Boysen et al. 2019). The selected system also needs a feasible, efficient, and comprehensive application strategy that should cover potential decisions regarding the implementation and operation of mobile robot systems.

The literature on mobile robots, mobile robot systems, and managerial decision areas in the warehouse context is fragmented (Table 1.1). Since other reviews only partially identify applicable mobile robot systems and fail to deliver a holistic managerial decision framework for their warehouse implementations, reviewing these aspects is well justified. In addressing the gaps mentioned above, this study identifies mobile robot systems adopted in warehouse operations and synthesises these systems' advantages and disadvantages. This book contributes to knowledge with a clear typology of mobile robots in warehouse operations. Moreover, it evaluates previous empirical works to highlight potential future research areas of

Table 1.1 Literature on mobile robots

Review paper	Review focus	Review outcome	Analysed mobile robot managerial decisions												
			Criteria for system evaluation	Identifying KPIs	Type and coordination of robots	Facility layout	Human-robot interaction	Storage assignment	Order management	Quantity of robots	Maintenance and failure handling	Energy management	Task allocation	Path planning	Conflict management
Vis (2006)	Planning and control of AGV systems	Focus areas to plan and control large AGV systems				✓				✓		✓	✓	✓	✓
Le-Anh and De Koster (2006)	Planning and control of AGV systems	A framework for design and control of AGV systems		✓	✓	✓				✓		✓	✓	✓	✓
Fazlollahtabar and Saidi-Mehrabad (2013)	AGV system path planning and task allocation algorithms	Methodology and algorithm categorisation											✓	✓	✓
Bechtsis et al. (2017)	Role of AGVs and AMRs in smart distribution and manufacturing systems	A framework covering key decisions with a sustainability perspective		✓	✓		✓						✓		
Azadeh et al. (2019)	Modelling, design, and control of automated order picking systems in warehouses	System analysis, design optimisation, and operations planning and control for robotic systems	✓					✓					✓		

(continued)

Table 1.1 (continued)

Review paper	Review focus	Review outcome	Analysed mobile robot managerial decisions												
			Criteria for system evaluation	Identifying KPIs	Type and coordination of robots	Facility layout	Human-robot interaction	Storage assignment	Order management	Quantity of robots	Maintenance and failure handling	Energy management	Task allocation	Path planning	Conflict management
Wior et al. (2018)	Automated transportation system types and influences of their interruptions	Advantages and disadvantages of transportation systems in various domains	✓	✓							✓				✓
Boysen et al. (2019)	Automated order picking systems in warehouses with an e-commerce focus	Analysis of automated order picking systems that suit B2C e-commerce	✓					✓	✓						
Jaghbeer et al. (2020)	Automation of order picking systems and their design implications	Performance of automated order picking systems and links between design and performance		✓			✓	✓	✓						
Fottner et al. (2021)	Enabling technologies and methods For autonomous intralogistics systems	A classification framework covering different automation stages for different intralogistics systems			✓					✓			✓	✓	✓

Reference	Focus	Contribution										
Fragapane et al. (2021)	Planning and control of AMRs	Technological developments and decision areas for the planning and control of AMRs		✓	✓	✓	✓	✓	✓	✓	✓	✓
This book	Planning and control of automated and autonomous systems in warehouses that include mobile robots	A managerial decision framework tailored for mobile robot implementation in warehouses	✓	✓	✓	✓	✓	✓	✓	✓	✓	✓

Source: Authors

mobile robot automation in warehouses. Finally, it develops a framework covering managerial decisions to be made before implementing mobile robot systems. This framework increases the probability of implementing suitable systems by considering system-specific characteristics. The following section outlines the systematic literature review (SLR) process. Then, with a clear mobile robot typology, systems adopted in warehouses are identified and analysed according to criteria to aid decision-makers in choosing the correct solution. The following section develops a framework of managerial decisions to consider before and during mobile robot systems implementation. Finally, the study evaluates previous empirical works to highlight potential future mobile robot automation research areas in warehouses.

REFERENCES

50,000 Warehouses to Use Robots by 2025 as Barriers to Entry Fall and AI Innovation Accelerates. (2019). ABI Research. https://www.abiresearch.com/press/50000-warehouses-use-robots-2025-barriers-entry-fall-and-ai-innovation-accelerates/

Adobe Holiday Shopping Trends. (2019). Adobe Analytics. https://www.adobe.com/content/dam/www/us/en/experience-cloud/digital-insights/pdfs/adobe_analytics-holiday-predictions-2019.pdf?promoid=NV3KR4X2&mv=other

Azadeh, K., De Koster, R. B. M., & Roy, D. (2019). Robotized and automated warehouse systems: Review and recent developments. In *Transportation Science* (Vol. 53, Issue 4). https://doi.org/10.1287/trsc.2018.0873

Bartholdi, J. J., & Hackman, S. T. (2019). Warehouse & Distribution Science. In *Available on line at: http://www.tli.gatech.edu…* (Issue August).

Bechtsis, D., Tsolakis, N., Vlachos, D., & Iakovou, E. (2017). Sustainable supply chain management in the digitalisation era: The impact of Automated Guided Vehicles. *Journal of Cleaner Production, 142,* 3970–3984. https://doi.org/10.1016/j.jclepro.2016.10.057

Bogue, R. (2016). Growth in e-commerce boosts innovation in the warehouse robot market. *Industrial Robot, 43*(6), 583–587. https://doi.org/10.1108/IR-07-2016-0194

Boysen, N., De Koster, R. B. M., & Weidinger, F. (2019). Warehousing in the e-commerce era: A survey. In *European Journal of Operational Research* (Vol. 277, Issue 2, pp. 396–411). Elsevier B.V. https://doi.org/10.1016/j.ejor.2018.08.023

Cambridge Dictionary. (n.d.). Cambridge Dictionary. Retrieved February 21, 2021, from https://dictionary.cambridge.org/dictionary/english/robot

De Koster, R. B. M. (2018). Automated and robotic warehouses: developments and research opportunities. *Logistics and Transport*, 2(38), 33–40. https://doi.org/10.26411/83-1734-2015-2-38-4-18

De Koster, R. B. M., Le-Duc, T., & Roodbergen, K. J. (2007). Design and control of warehouse order picking: A literature review. In *European Journal of Operational Research* (Vol. 182, Issue 2). https://doi.org/10.1016/j.ejor.2006.07.009

Fazlollahtabar, H., & Saidi-Mehrabad, M. (2013). Methodologies to Optimize Automated Guided Vehicle Scheduling and Routing Problems: A Review Study. *Journal of Intelligent and Robotic Systems: Theory and Applications*, 77(3–4), 525–545. https://doi.org/10.1007/s10846-013-0003-8

Fottner, J., Clauer, D., Hormes, F., Freitag, M., Beinke, T., Overmeyer, L., Gottwald, S. N., Elbert, R., Sarnow, T., Schmidt, T., Reith, K. B., Zadek, H., & Thomas, F. (2021). Autonomous systems in intralogistics-state of the art and future research challenges. *Logistics Research*, 14(1). https://doi.org/10.23773/2021_2

Fragapane, G., de Koster, R., Sgarbossa, F., & Strandhagen, J. O. (2021). Planning and control of autonomous mobile robots for intralogistics: Literature review and research agenda. *European Journal of Operational Research*, 294(2), 405–426. https://doi.org/10.1016/j.ejor.2021.01.019

Frazelle, E. H. (2016). *World-Class Warehousing And Material Handling* (Second Edi). McGraw-Hill Education.

Horňáková, N., Jurík, L., Hrablik Chovanová, H., Cagáňová, D., & Babčanová, D. (2019). AHP method application in selection of appropriate material handling equipment in selected industrial enterprise. *Wireless Networks*. https://doi.org/10.1007/s11276-019-02050-2

Huang, G. Q., Chen, M. Z. Q., & Pan, J. (2015). Robotics in ecommerce logistics. *HKIE Transactions Hong Kong Institution of Engineers*, 22(2), 68–77. https://doi.org/10.1080/1023697X.2015.1043960

Jaghbeer, Y., Hanson, R., & Johansson, M. I. (2020). Automated order picking systems and the links between design and performance: a systematic literature review. *International Journal of Production Research*, 0(0), 4489–4505. https://doi.org/10.1080/00207543.2020.1788734

Kattepur, A., Rath, H. K., Simha, A., & Mukherjee, A. (2018). Distributed optimization in multi-agent robotics for industry 4.0 warehouses. *Proceedings of the ACM Symposium on Applied Computing*, 808–815. https://doi.org/10.1145/3167132.3167221

Le-Anh, T., & De Koster, R. B. M. (2006). A review of design and control of automated guided vehicle systems. *European Journal of Operational Research*, 171(1), 1–23. https://doi.org/10.1016/j.ejor.2005.01.036

Rushton, A., Croucher, P., & Baker, P. (2014). *Handbook of logistics and distribution management*.

Vis, I. F. A. (2006). Survey of research in the design and control of automated guided vehicle systems. *European Journal of Operational Research*, *170*(3), 677–709. https://doi.org/10.1016/j.ejor.2004.09.020

Wior, I., Jerenz, S., & Fay, A. (2018). Automated transportation systems subject to interruptions in production and intralogistics - a survey and evaluation. *International Journal of Logistics Systems and Management*, *30*(4), 421. https://doi.org/10.1504/IJLSM.2018.093581

Wurman, P. R., DAndrea, R., & Mountz, M. (2008). *Coordinating Hundreds of Cooperative, Autonomous Vehicles in Warehouses.*

CHAPTER 2

Methodology

The content prepared for this book is obtained through what is called a systematic literature review (SLR). The rationale behind the SLR is to have a clear view of mobile robot system applications in the literature and the complications faced while applying those systems to warehouses. The SLR summarises the existing information by being thorough and unbiased (Denyer and Tranfield 2009) and reproducible. Compared to a traditional literature review, the SLR is robust, scientific, and transparent (Tranfield et al. 2003). The research protocol followed in this review was adopted from Tranfield et al. (2003) and comprises two main stages.

2.1   PLANNING THE REVIEW

As the first stage of Tranfield et al. (2003) recommends, a scoping study was carried out together with the review panel consisting of three writers, two academicians, and a library expert. The scoping study aimed to be able to explore the literature broadly to devise solid review questions.

The literature on mobile robot systems applied in warehouse operations as an automation solution and their managerial decision aspects is covered in this case. This topic includes two review questions:

RQ1. How are mobile robot systems applied in warehousing operations?

A. Yildirim et al., *Mobile Robot Automation in Warehouses*, Palgrave Studies in Logistics and Supply Chain Management, https://doi.org/10.1007/978-3-031-12307-8_2

RQ2. How can managerial decisions be structured when adopting mobile robot systems in warehouses?

A list of relevance criteria was developed to guide the literature selection. Only articles from the year 2000 and conference papers from 2015 to date were considered relevant owing to the recency of the technology. All papers reviewed are published in English. Papers were expected to be peer-reviewed, and their full text was expected to be accessible. Finally, only the papers that are about warehouses and mobile robots were selected in line with the review context (Table 2.1).

Quality criteria were adopted from Pittaway et al. (2004) to select and evaluate which publications to include in the review. This study's four quality criteria are theory robustness, contribution to knowledge, methodology and arguments, and implications for practice (Table 2.2).

Table 2.1   Selection criteria for the SLR

Selection criteria	Inclusion	Explanation
Language	Papers should be in English	It is the language of this book
Accessibility	The full text of the papers should be accessible	Papers should be available for evaluation
Review	Peer-reviewed papers	A certain level of quality and validity of the papers should be ensured
Type of publication	Academic journals and conference papers	To keep the quality high and the number of papers manageable
Year of publication	Academic journals published later than 2000 or conference papers/proceedings published later than 2015	Mobile robot systems in warehousing drastically developed after the year 2000, so papers before 2000 would miss breakthroughs. Quality conference papers older than five years are assumed to have become academic journals; thus, they are excluded
Scope and context	Mobile robot systems in warehouses will be considered	Other contexts, such as manufacturing and systems that do not involve mobile robots, are not in the scope of this research

Source: Authors

Table 2.2 Quality criteria for the SLR

Quality criteria	Theory robustness	Contribution to knowledge	Methodology and arguments	Implication for practice
1	The literature review is weak or does not exist. Even though a theory exists, it is not supported effectively	The contribution is either not advanced or not precise	The logic behind the data is not supported, or the methodology is not robust	The concepts and ideas presented are ambiguous or somewhat irrelevant; thus, they are challenging to implement
2	The theory is somewhat validated. A basic level of literature review supports the topic-theory-data match	Contribution builds upon the existing ideas or studies	The logic behind data is supported, but it is limited, or methodology/ research design could be improved	There is a potential for implementing the proposed ideas with minor revisions or adjustments
3	The theory is nicely explained and supported by the presence of a relevant literature review. The approach fits the topic and the data	The paper expands the issue with an innovative approach and well-explained solution	Ideas well support data, the research design is robust, and the analysis is rigorous	A significant benefit may be obtained if the ideas being discussed are put into practice
N/A	This criterion is not applicable	This criterion is not applicable	This criterion is not applicable	This criterion is not applicable

Adopted from Pittaway et al. (2004)

2.2 Conducting the Review

Two research contexts are defined as 'warehouse' and 'mobile robots' to answer the above-presented review questions in the scoping study. Two keyword groups were formed as broad as possible for both contexts. Out of those keyword groups, keyword strings were gathered, and these strings were named as 'S1' and 'S2' for 'warehouse' and 'mobile robots', respectively. These search strings were then used to search databases with the 'AND' operator.

The selected databases are Scopus, Web of Science, ABI Inform, and EBSCO, as they include most of the relevant journals. Database search

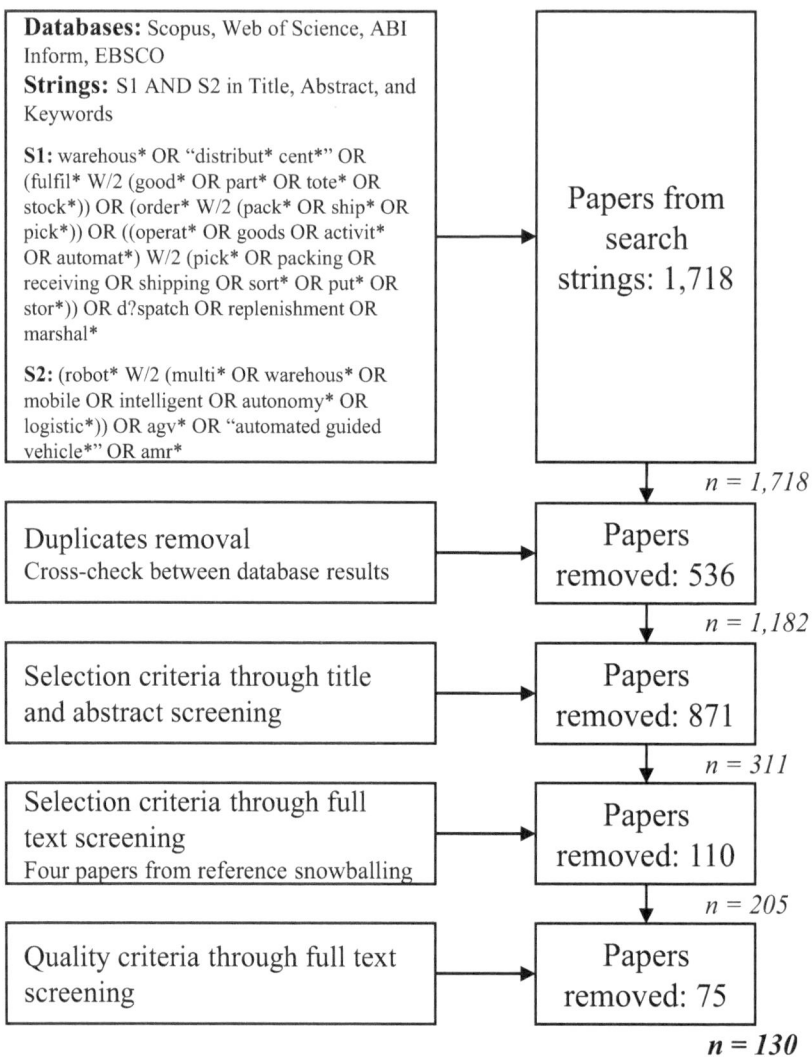

Databases: Scopus, Web of Science, ABI Inform, EBSCO
Strings: S1 AND S2 in Title, Abstract, and Keywords

S1: warehous* OR "distribut* cent*" OR (fulfil* W/2 (good* OR part* OR tote* OR stock*)) OR (order* W/2 (pack* OR ship* OR pick*)) OR ((operat* OR goods OR activit* OR automat*) W/2 (pick* OR packing OR receiving OR shipping OR sort* OR put* OR stor*)) OR d?spatch OR replenishment OR marshal*

S2: (robot* W/2 (multi* OR warehous* OR mobile OR intelligent OR autonomy* OR logistic*)) OR agv* OR "automated guided vehicle*" OR amr*

Papers from search strings: 1,718

n = 1,718

Duplicates removal
Cross-check between database results

Papers removed: 536

n = 1,182

Selection criteria through title and abstract screening

Papers removed: 871

n = 311

Selection criteria through full text screening
Four papers from reference snowballing

Papers removed: 110

n = 205

Quality criteria through full text screening

Papers removed: 75

n = 130

Fig. 2.1   Steps of the SLR. (Source: Authors)

results were exported to MS Excel, and the duplicates were removed. Titles and abstracts of 1182 papers were screened for possible consideration of full texts according to the relevance criteria identifying 315 articles, including four papers identified through cross-references (snow-balling). The remaining papers were re-evaluated through full-text reading and using the criteria in Table 2.2, leaving 205 papers for quality appraisal.

In the quality appraisal stage, papers were rated on a scale of zero to three in each of the four criteria, and papers scoring 8 out of 12 were qualified for review. In total, 130 papers were selected for review (Fig. 2.1). One hundred twelve papers are categorised as 'empirical' papers, whereas eighteen are 'review' papers.

References

Denyer, D., & Tranfield, D. (2009). Producing a systematic review. In D. A. Buchanan & A. Bryman (Eds.), The SAGE handbook of organizational research methods. *The Sage Handbook of Organizational Research Methods*, 671–689.

Pittaway, L., Robertson, M., Munir, K., Denyer, D., & Neely, A. (2004). Networking and innovation: A systematic review of the evidence. *International Journal of Management Reviews, 5–6*(3–4), 137–168. https://doi.org/10.1111/j.1460-8545.2004.00101.x

Tranfield, D., Denyer, D., & Smart, P. (2003). Towards a Methodology for Developing Evidence-Informed Management Knowledge by Means of Systematic Review* Introduction: the need for an evidence-informed approach. *British Journal of Management, 14*, 207–222.

CHAPTER 3

Mobile Robot Systems and Their Evaluation

Systems emerging from AGVs and AMRs could be categorised in a few ways, such as their type of configuration (Singhal et al. 2018), their type of interaction (Kattepur et al. 2018; Singhal et al. 2018), or their type of navigation (Wior et al. 2018). Mentioning these systems in the navigation context is preferred in this book to be able to precisely separate the systems because systems could fit in multiple categories in other types of categorisations. For instance, in configuration categorisation of Singhal et al. (2018), mobile robots in a specific system could be controlled via centralised, decentralised, or distributed configurations. As mobile robots in that system could be arranged to share the same goal or have their own goals, a mobile robot system could be listed in both cooperative and collaborative interaction. In cooperative interaction, there is a system goal to be achieved by working together, whereas in collaborative interaction, there are independent objectives that might even cause conflicts between robots.

Wior et al. (2018) categorise automated transport systems with mobile robots into three according to their types of navigation: rail using, guided, and freeway. This review adopts this classification and defines navigation types in four main categories: linear route, guided, and freeway mobile robots, as well as hybrid systems. Linear route mobile robots cover rail-using robots of Wior et al. (2018) and include wire-using robots to extend the coverage as linear route mobile robots. Furthermore, hybrid systems

© The Author(s), under exclusive license to Springer Nature   17
Switzerland AG 2023
A. Yildirim et al., *Mobile Robot Automation in Warehouses*, Palgrave
Studies in Logistics and Supply Chain Management,
https://doi.org/10.1007/978-3-031-12307-8_3

are included as the fourth navigation type since there is a natural opportunity to combine those systems among themselves or with other systems such as Automated Storage and Retrieval Systems (AS/RS). Sub-systems under these main categories and the warehouse operation context they were studied are presented in Table 3.1. Details of these papers can be found in the Appendix. Note that in many papers, two operations or two sub-systems are mentioned simultaneously.

Systems such as robot-based compact storage and retrieval systems or autonomous vehicle-based storage and retrieval systems are left out of the scope of this study because mobile robots in these systems are a part of the main structure, and they cannot act in the absence of those structures.

Schmidt and Schulze (2009) state four well-known criteria for evaluating mobile robot systems: cost, service-level, flexibility, and scalability. Apart from the service-level criterion, the other three criteria are also mentioned in many other studies (Azadeh et al. 2019; Bauters et al. 2016; Huang et al. 2015; Roy et al. 2019; Zou et al. 2018; Hanson et al. 2018). Firstly, to prevent the cost criterion from becoming too broad, this paper only considers mobile robot costs directly connected to capability requirements, such as onboard intelligence and gripping arms. Moreover, flexibility is mentioned in two different ways: 'Flexibility of the Infrastructure' and 'Flexibility in Material Handling'. Scalability is simply the ability of mobile robot systems to cope with demand fluctuations. Finally, according to Benady (2016), 80% of the warehouses are manually operated with no automation. Once they want to change their operational model, they will need to consider the required implementation time of the mobile robot systems as a decision criterion. For this reason, the 'Time to Implement' criterion is also considered in this study to give the managers an idea of applicable systems in terms of the time they require to initiate operations. Explanations of these criteria in terms of ratings are listed in Table 3.2— '3' indicates relatively high performance in the mentioned criterion, whereas '1' indicates relatively low performance.

Considering these factors, an overview of mobile robot systems is made (Table 3.3). These systems and their use cases in warehouses are represented in the following sub-sections using real-life case examples, and also by considering the six factors described above. The authors also utilised additional selected sources to supplement the information available in the reviewed academic literature.

Table 3.1 Mobile robot systems and warehouse operations

Operation/ sub-system	Linear route mobile robots		Guided mobile robots		Freeway mobile robots		Hybrid systems			
	Rail	Wire	Barcode-guided robots	Laser-guided robots	Autonomous forklifts	Human-collaborated robots	Mobile picking robots	AS/RS, conveyors, mobile robots	Picker and transport robots	Laser-guided robots and pallet shuttles
Picking	1	0	39	0	2	3	7	1	2	1
Replenishment	0	0	3	0	0	0	0	0	0	0
Put-away	0	0	0	0	2	0	0	1	0	1
Sorting	0	0	1	0	0	0	0	1	0	0
Shipping	0	0	0	0	4	0	0	0	0	0
Not mentioned	0	0	2	3	4	1	0	0	0	0
Sub-system paper totals	1	0	43	3	10	4	7	2	2	1
No sub-system mentioning papers		1	3		15				0	
Main system paper totals		**2**	**49**		**36**				**5**	
No main system mentioning paper totals	**38**									
Total papers					**130**					

Source: Authors

Table 3.2   Mobile robot evaluation criteria and ratings

Rating	Mobile robot cost	Flexibility of the infrastructure	Flexibility in material handling	Scalability	Time to implement
3	The robot does not have onboard intelligence and handling function (forks and gripping or vacuuming arms)	System infrastructure is not fixed; it is easy to adapt the system to process and/or layout changes	The system is capable of handling bulky products (palletised or similar) or small products (bins/ boxes/totes)	The system responds to dynamic demand changes that exceed capacity limits within days (with additional robots)	The system needs weeks for a complete set-up (mobile robots, layout, and software installation)
2	The robot either has onboard intelligence or a handling function	System infrastructure is not entirely fixed; it can adapt to process and/or layout changes but requires a new set-up	The system is capable of handling many types of products but might require adjustments or configurations.	The system requires a few weeks to respond (additional robots, low-level construction, or human workforce requirements)	A few months is necessary for a complete set-up (railways, barcodes, or laser system installations that partly change the warehouse layout)
1	The robot has both onboard intelligence and handling function	Infrastructure is fixed, and it is not suitable to adapt to process and/or layout changes	The system can only operate with either bulky products (palletised or similar) or small products (bins/ boxes/totes)	The system requires several weeks to respond (high-level construction requirements)	Several months are necessary for a complete set-up (high-level construction that changes the whole warehouse layout)

Source: Authors

Table 3.3   Mobile robot system evaluations

Mobile robot systems (system context)	Mobile robot cost	Flexibility of infrastructure	Flexibility in handling	Scalability	Time to implement
Rail- or wire-using robots (P2G)	3	1	3	2	2
Barcode-guided robots (G2P)	2	2	3	2	2
Laser-guided robots (R2G)	2	2	1	2	2
Autonomous forklifts (R2G)	1	3	1	3	3
Human-collaborated robots/co-bots (P2G)	2	3	2	2	3
Mobile picking robots (R2G)	1	3	1	3	3
AS/RS, conveyors, linear route robots (R2G)	3	1	1	1	1
Picker and transport robots (R2G)	1	3	1	3	3
Laser-guided robots and pallet shuttles (R2G)	3	1	1	1	1

Source: Authors

3.1   LINEAR ROUTE MOBILE ROBOTS

Mobile robots that follow a linear route made of wires, magnetic tracks or rails have been present in logistics warehouses to transport heavy products within the warehouse since the 1950s (Wurman et al. 2008). Even though these systems have predetermined navigation, they are still preferred in warehouses as they can be easily controlled and hardly damaged, making them highly reliable (Wior et al. 2018).

Recent P2G examples such as automated trolleys are also adaptable to chaotic warehouse contexts (Füßler et al. 2019). In this system, a human picker travels together with the trolley through the warehouse environment, and once the trolley approaches an ordered product on the way, it stops and shows the product through its screen. This system is generally preferred in case picking or medium to large products. It is suitable for picking up to 450,000 cases per day, increasing picking performance by 40% compared to manual picking (Swisslog n.d.-b).

Linear route robots can also be used in put-away operations by carrying materials to an AS/RS or a simple shelf (Liu 2018). Furthermore, they can be used in sorting operations together with a conveyor by carrying the batched boxes to the shipping area (Abbas et al. 2018).

Although, to our knowledge, no paper has mentioned linear route mobile robots in a chaotic warehouse setting, they are used in e-commerce order picking operations in warehouses in a G2P context. In this setting, products could be put on totes that are above the AGVs (Fig. 3.1), and they could transport them to packing stations. These robots can carry products up to 35 kg and can also be used in a multi-floor warehouse (SSI Schaefer n.d.-a).

Fig. 3.1   AGV weasel. (Source: SSI Schaefer (n.d.-a))

There are two types of linear route mobile robots: rail-using robots and wire-using robots.

In the rail-using robot version, a rail route is constructed to the warehouse floor as guidance, and mobile robots travel along the rails to carry products. Railways on the floor could pose a safety issue which would require careful maintenance (Schulze and Zhao 2007).

In the wire-using robots version, robots travel on a wired floor which is continuously energised, and they get the inductive power generated from the magnetic field over the floor (Schulze and Zhao 2007). For this reason, this system is the only system that does not require refuelling to travel, eliminating potential robot-charging problems. Furthermore, they do not need rails to travel, which prevents the safety issues that could be caused by parts like screws or pins falling from the railway (Schulze and Zhao 2007).

As rail-using and wire-using robots have similar outcomes in the criteria evaluation, one system evaluation would be sufficient.

Mobile Robot Cost: Linear route mobile robots cannot move freely in the environment; therefore, they cannot bypass an obstacle. Thus, instead of AMRs, AGVs are generally preferred for such systems that do not require extensive onboard intelligence (Wior et al. 2018) (Rating: 3).

Flexibility of the Infrastructure: The infrastructure (railways or wires) they stand on is fixed. For this reason, these systems are not adaptable to process or layout changes (Rating: 1).

Flexibility in Material Handling: Being able to carry bulky products on pallets or small products in tote bins makes them flexible in the types of products they can carry (Rating: 3).

Scalability: They are not perfectly scalable as they require new routes and low-medium level of construction (floors), which might last for weeks to increase the system capacity (Rating: 2).

Time to Implement: SSI Schaefer (n.d.-b) mentions the implementation time of the system as five weeks, which also suits our rating table since these systems require a low-medium level of construction that partly changes the warehouse floors for installation (Rating: 2).

3.2   Guided Mobile Robots

Guided mobile robots can move non-linearly and are, therefore, preferable compared to linear route mobile robots in terms of navigation flexibility. However, they should be directed through barcodes or laser beams scattered in the warehouse environment, and they cannot leave these specific environments to navigate freely.

There are two main types of guided robots: barcode-guided robots and laser-guided robots.

3.2.1   Barcode-Guided Mobile Robots

Barcode-guided mobile robots became famous with the acquisition of Kiva Systems by Amazon. After this acquisition, Amazon increased its worker productivity up to threefold and inspired many companies to adopt similar systems (Enright and Wurman 2011). Mobile robots in these systems travel on grid-based floor layouts in a G2P context, and they cannot leave these grids. Barcode guidance is generally used in order picking operations with shelf-carrying robots, but it also inspired solutions to other warehouse operations such as sortation in which robots carry packages (Fan et al. 2018; Liu et al. 2019).

Shelf-carrying robots operate in two types of order picking set-up: item fetch and order fetch. Both versions are in a G2P context. Shelves are aligned in three-foot squares in the item fetch system, each square having barcodes in the middle. The mobile robots carry the necessary shelf to the packing station, where a human picker stands, ready to pick up the required items. These items are then put into the packages by the picker and placed in a conveyor to be transported to the shipping station (Enright and Wurman 2011; Wurman et al. 2008). If the set-up is order fetch, additional robots get into the area and carry picked and packed boxes to the shipping stations. In other words, the conveyor system is eliminated with the presence of extra mobile robots (Enright and Wurman 2011; Yoshitake et al. 2019). The layout of the typical shelf-carrying mobile robots is presented in Fig. 3.2. Geek+ has many applications that increase human pickers' productivity to around 300 pieces/hour/station, up from approximately 100 pieces/hour/station (Geek+ n.d.-c, n.d.-d). Likewise, Hikrobotic's barcode-guided robots have reportedly increased DHL Express' worker efficiency by 33% and space utilisation by 40% (Hikrobotics n.d.-a).

Due to the increase in demand in e-commerce, manual sortation of parcels has become a time-consuming activity, and using package-carrying mobile robots for sortation has also gained attention (Fan et al. 2018). Funnels are placed on a grid layout to implement the system, and mobile robots are expected to take the boxes/parcels that are coming from the

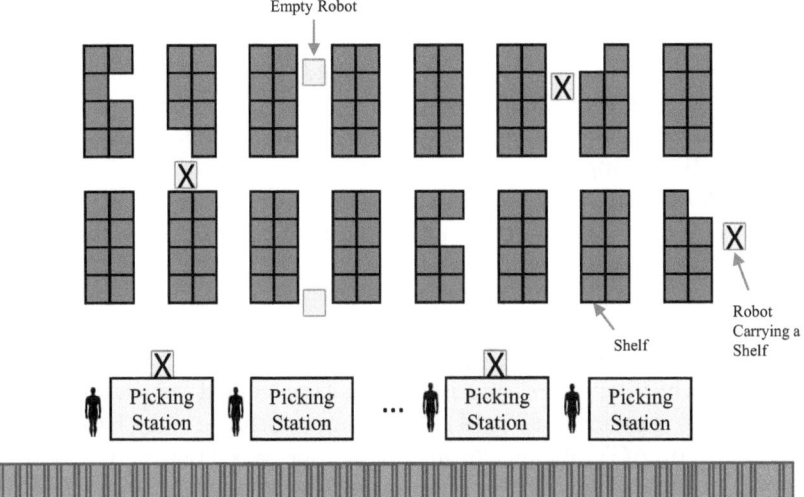

Fig. 3.2 Typical layout for shelf-carrying mobile robots. (Reproduced from Yuan et al. (2019))

packing station through a conveyor and drop these boxes into the relevant funnel (Liu et al. 2019). Instead of funnels, distribution conveyors can also be used at the edges of the grid area. Using ten mobile robots next to a worker, sortation throughput can be increased by up to tenfold, from 60 parcel/hour/worker to 600 parcel/hour/worker (Prime Vision n.d.).

Mobile Robot Cost: AGVs are preferred to AMRs as this system does not necessitate freeway navigation. However, as the number of drive units increases, AMRs are preferable since it gets computationally hard to control and direct hundreds of drive units from a central control tower (Rating: 2).

Flexibility of the Infrastructure: They require a dedicated area that might require additional time to adapt to the potential process/layout changes (Wurman et al. 2008) (Rating: 2).

Flexibility in Material Handling: They can carry shelves with tote bins to pickers or boxes to sortation stations, making these systems flexible in terms of the type of products they can carry (Enright and Wurman 2011; Liu et al. 2019). Besides, they can carry palletised products with system adjustments as they can carry up to 1300 kgs of weight (Azadeh et al. 2019) (Rating: 3).

Scalability: According to Wurman et al. (2008), these systems are typically formed of hundreds of robots, and adding more robots, shelves, and packing stations (minor construction) would increase the capacity of the warehouse. Likewise, removing them temporarily would decrease the capacity, making these systems preferable for varying workloads (Boysen et al. 2019) (Rating: 2).

Time to Implement: They require barcode installation and unique rack placement in a dedicated area which requires a few months (Geek+ n.d.-b) (Rating: 2).

3.2.2   Laser-Guided Mobile Robots

There has been an increasing interest in laser-guided vehicles since the early 2000s (Ferrara et al. 2014). According to Schulze and Zhao (2007), one-third of AGV manufacturing in Europe was laser-guided mobile robots in 2005. They only require mirrors in strategic locations to help them map and navigate freely across that environment in an R2G setting (Ferrara et al. 2014; Ly 2019). Twenty of these robots can increase the throughput times by 30% compared to manual forklifts (Swisslog n.d.-a). Laser-guided mobile robots are designed to carry pallets or roll cages (i.e. forklifts and reach trucks), preventing human errors (Ferrara et al. 2014; Ly 2019; Polten and Emde 2020). For this reason, they are generally being used in replenishment and put-away rather than order picking operations in the warehouse unless the orders are not received as full pallets.

These systems might be dangerous when they operate in the same environment as human workers. Thus, human workers generally get used to their travelling paths and leave the robot zone free, preventing them from working together and ensuring safety (Raineri et al. 2019).

Mobile Robot Cost: AGVs with forks are generally preferred for laser-guided operations as they are sufficient for predefined path navigation (Fig. 3.3) (Rating: 2).

Flexibility of the Infrastructure: These solutions take advantage of mirrors placed in an operational area so they can adapt to a new process/layout with a new set-up (Rating: 2).

Flexibility in Material Handling: Laser-guided mobile robots can carry only pallets (Rating: 1).

Scalability: They are scalable, just like barcode-guided mobile robots, since adding more robots would solely be enough to increase the system's capacity. However, they are large vehicles, and adding too many of them

Fig. 3.3 Laser-guided robots. (Source: Hikrobotics (n.d.-b))

could consume a lot of warehouse space (Polten and Emde 2020). Further, increasing the number of robots might necessitate using extra mirrors as an extra set-up time (Rating: 2).

Time to Implement: Installation of mirrors and the integration of the system could take up to several months even though it would mean low-medium-level construction (Banker 2018) (Rating: 2).

3.3 FREEWAY MOBILE ROBOTS

Freeway mobile robots do not require any construction and can move freely in the warehouse environment autonomously. Generally, AMRs are preferred in these types of systems due to their ability to map the warehouse, recalculate the path, and bypass the obstacles autonomously with their onboard intelligence.

On the other hand, having plenty of movement possibilities results in complex and computationally expensive path planning algorithms (Wior et al. 2018). Moreover, free navigation could cause safety issues since they might need to work with human workers in the same environment.

There are three types of freeway mobile robot systems; autonomous forklifts, human collaborated mobile robots (co-bots), and mobile picking robots.

3.3.1   Autonomous Forklifts

Having similar characteristics with laser-guided robots, autonomous fork-lifts mainly carry pallets in the warehouse in an R2G context (Abbas et al. 2018; Polten and Emde 2020). For that reason, in warehouses with many orders and small-medium product sizes, they are generally not preferred in order picking activities. Instead, they are used in receiving (unloading), put-away, replenishment, and shipping (loading) activities (Draganjac et al. 2016; Polten and Emde 2020). According to DHL, autonomous forklifts increased the throughput by 20%, and they have more than 98% of reliability (DHL n.d.-a). They also significantly improve the safety factor in pallet handling, one of the considerably dangerous activities in the warehouse when handled by human workers.

Mobile Robot Cost: Although they might not be as expensive as a gripping autonomous mobile robot, these AMRs also possess forks that move and lift pallets autonomously. According to Horňáková et al. (2019), each autonomous forklift costs $60,000 (Rating: 1).

Flexibility of the Infrastructure: They do not have any fixed infrastructures, making them highly flexible to adapt to different processes or layouts (Rating: 3).

Flexibility in Material Handling: They cannot handle tote bins or boxes as they are designed to transport pallets (Rating: 1).

Scalability: Even though they are oversized vehicles like laser-guided robots, they are suitable for capacity adjustment since they can scale up or down with more or less mobile robots (Rating: 2).

Time to Implement: They have a relatively short implementation time as they do not require construction and are highly adaptable to any warehouse layouts (Geek+ n.d.-a) (Rating: 2).

3.3.2   Human-Collaborated Mobile Robots

Human-collaborated mobile robots (co-bots) are used in order to pick and operate with human pickers in a P2G context. There are two alternative scenarios where human collaborated mobile robots could be applied: fixed-assigned and free-floating. In the fixed-assigned scenario, the robot travels with a human picker. The picker picks the products from the pick-list and puts them onto the robot. Once the robot reaches the capacity or the order list is complete—whichever is earlier—it autonomously returns to the packing station and another robot is requested to meet the picker.

In the free-floating scenario, human pickers stand in the aisles of the warehouse, and mobile robots travel in the warehouse according to their picking list. Once they request a product, they travel to the correct aisle and wait for the picker to load the product. When they reach their capacity, they return to the packing station and unload the products (Boysen et al. 2019).

These scenarios can double the worker productivity and orders processed while reducing the worker training time by up to 80% (Expeditors n.d.; Locus Robotics n.d.-a).

On the contrary, mobile robots and human pickers should be optimised jointly to optimise the picking operation (Boysen et al. 2019). Moreover, mobile robots and human pickers should operate closely in the same environment, increasing the probability of having safety issues (Petković et al. 2019; Raineri et al. 2019).

Mobile Robot Cost: Although closely following a human worker necessitates AMRs, gripping functionality is not required (Rating: 2).

Flexibility of the Infrastructure: These robots do not have any fixed infrastructure, so they are adaptable to layout or process adjustments (Rating: 3).

Flexibility in Material Handling: They can handle small products or boxes/bins/totes, but they cannot operate with bulky palletised products as human workers cannot lift them (Rating: 2).

Scalability: They are suitable for a chaotic warehouse setting since they can coordinate and scale with more mobile robots, but they also require an additional human workforce (Rating: 2).

Time to Implement: They have a short implementation time, ranging from days to a few weeks because they can start working as soon as they are placed in the warehouse (Locus Robotics n.d.-b). The human-collaborated mobile robot system is a highly adaptive system which does not require a re-layout while transforming manually operated order picking in conventional warehouses (De Koster 2018; Wang et al. 2019) (Rating: 3).

3.3.3   Mobile Picking Robots

Mobile picking robots are relatively new and popular because they can do the picking on their own in the R2G context. This feature removes human errors and human costs in the order picking operation (Fig. 3.4).

Fig. 3.4   Mobile picking robots. (Source: Magazino (n.d.))

Mobile picking robots generally have compartments or tote bins on them, and they put the products they pick into these spaces. Once they reach their capacity, they return to the packing area to unload (Bogue 2016; Huang et al. 2015). Only three of these robots could save 40 hours a week, which is close to a full-time worker (MiR n.d.). Eight robots put 2500 pieces away in another set-up and pick 700 pieces every hour while storing 1.8 times more products in the same environment (Hai Robotics 2019).

On the other hand, robot picking is still inferior to human picking in terms of speed and accuracy (Wang et al. 2019). Even though they now lack speed and accuracy, researchers and practitioners focus on the topic of investigating new technologies by running contests such as Amazon Picking Challenge (Bormann et al. 2019).

Mobile Robot Cost: This system requires the robot to have a gripping functionality and the high-level onboard intelligence, which increases robots' costs (Rating: 1).

Flexibility of the Infrastructure: Like human-collaborated mobile robots, these robots do not have any fixed infrastructure, so they are adaptable to layout or process adjustments (Rating: 3).

Flexibility in Material Handling: They have limited flexibility in terms of small products they can carry because of products' weight (up to 6 kg) and shape (boxes or products that have edges) (Kimura et al. 2015; Magazino n.d.). Moreover, the picking areas of the robots should be structured for correct and efficient picking (Huang et al. 2015). Yet, there are ongoing object detection studies to identify products in an unstructured shelf correctly (Bormann et al. 2019). A stronger version of these robots can also operate with cases and pallets, but they cannot handle tiny and bulky products simultaneously (Hai Robotics 2019; MiR n.d.) (Rating: 1).

Scalability: They are suitable for a chaotic warehouse setting since they can coordinate and scale with more mobile robots without additional requirements (Rating: 3).

Time to Implement: They need weeks rather than months to implement as they become operational immediately upon installation (inVia Robotics n.d.) (Rating: 3).

3.4   HYBRID SYSTEMS

Hybrid systems combine at least one mobile robot system with other automation solutions. They might be preferable to remove the disadvantage of a specific system or to take advantage of a higher level of automation. The literature reveals three types of hybrid systems, all of which operate in the R2G method: (1) AS/RS, conveyors, and linear route robots; (2) picker and transport robots; (3) laser-guided robots and pallet shuttles.

3.4.1   *AS/RS, Conveyors, and Linear Mobile Robots*

In an R2G context, these systems are generally helpful if the warehouse traffic consists mainly of pallets or similar handling units according to the capability of the AS/RS. Even though there are other types of AS/RS claimed to be faster and that can store and retrieve small products (autonomous vehicle-based storage and retrieval systems), none of the papers mentions a hybrid system that includes mobile robots with different types of AS/RS (Azadeh et al. 2019).

In this solution, when the pallets arrive at the warehouse, they are put on a conveyor and transported to an area where mobile robots operate. Mobile robots transport these pallets from the conveyor to the AS/RS. The retrieval process is the reverse of the storage process (Liu 2018). These systems could automate many operations simultaneously, such as put-away, storage, and order picking, with a significantly lower risk of product damage compared to manual operations (Dematic n.d.). However, optimising one sub-system does not enhance overall efficiency. Thus, each system should be jointly optimised (Liu 2018). Moreover, these systems carry the disadvantages of each sub-system. For instance, AS/RS is not a scalable solution, and to scale it up, construction work is required, which could also be costly and time-consuming (Roy et al. 2019).

Mobile Robot Cost: AGVs suffice to carry the products or pallets that are taken from the AS/RS or conveyors (Rating: 3).

Flexibility of the Infrastructure: The system is fixed, and it is not adaptive to any process or layout changes due to its infrastructure (Rating: 1).

Flexibility in Material Handling: Together with having the capability of transporting different sizes of products, an AS/RS can only carry one type of handling unit in a set-up due to its physical build-up (Bauters et al. 2016; Wurman et al. 2008) (Rating: 1).

Scalability: The system is not scalable as it would need new and high-level construction to increase the system capacity (Roy et al. 2019; Wurman et al. 2008) (Rating: 1).

Time to Implement: It takes between six months and a year to construct the system to become operable (Cribley 2014) (Rating: 1).

3.4.2   Picker and Transport Robots

Picker and transport robot systems are similar to the human-collaborated system. Human pickers are replaced with picker robots (mobile picking robots) which can fetch and lift small products, whereas the transport robots are faster mobile robots that can carry products placed into tote bins or pallets on them in an R2G context (Fig. 3.5) (Bogue 2016; Lee and Murray 2019). This system can completely automate order picking operations (Kimura et al. 2015). However, it is noted that there are not many installations worldwide using this solution.

Fig. 3.5 Picker and transport robots. (Source: Robots (n.d.))

Mobile Robot Cost: This system requires both an AMR with the gripping functionality and an AMR as a transporter. For this reason, mobile robot costs are relatively high (Rating 1).

Flexibility of the Infrastructure: Like mobile picking robots, these robots do not have any fixed infrastructure, so they are adaptable to layout or process adjustments (Rating: 3).

Flexibility in Material Handling: This system is not flexible in product types or product handling units that can be carried. For instance, the picking robot of Fetch Robotics could lift products that weigh up to 6 kg, and its arm has a reach of 2 m (Bogue 2016) (Rating: 1).

Scalability: Even though the system requires two different types of robots, they are scalable (Huang et al. 2015) (Rating: 3).

Time to Implement: Similar to mobile picking robots, these systems are assumed to need weeks to implement as robots could operate as soon as they are placed in the warehouse with a structured picking area and require no construction (Huang et al. 2015; inVia Robotics n.d.) (Rating: 3).

3.4.3   Laser-Guided Mobile Robots and Pallet Shuttles

These systems differ from a laser-guided mobile robot system in automating pallet movements in shelves from the face to the interior with pallet shuttles in an R2G context (Fig. 3.6). Pallets are stored on rails, narrow

Fig. 3.6   Pallet shuttles and a forklift. (Source: Dexion (n.d.)))

aisles are removed, and pallet storing is optimised with the help of these pallet shuttles (Ferrara et al. 2014). On the other hand, laser-guided mobile robots cannot reach pallets directly. Instead, they need to pick up the pallet shuttle and put it in the appropriate rack for the pallet shuttle to carry the pallet to the face of the rack (Ferrara et al. 2014).

Even though this concept is only being mentioned in the literature, it is a feasible solution as pallet shuttles are commonly being used with human-operated forklifts. Yet, on top of laser-guided vehicles' benefits, pallet shuttle system promises to increase the storage density of the warehouse up to 130% by removing extra warehouse aisles and storing more pallets instead (Dexion n.d.).

Mobile Robot Cost: AGVs are generally preferred for these systems, similar to the case with laser-guided robot solutions (Rating: 3).

Flexibility of the Infrastructure: Shelving units are fixed infrastructures that are hardly adaptable to process or layout adjustments (Rating: 1).

Flexibility in Material Handling: This system only works with pallets or similar handling units, which decreases the level of flexibility in material handling (Rating: 1).

Scalability: There is high-level construction work for the shelves that hold pallet shuttles and light construction work due to the laser-guided robot system structure that prevents rapid capacity alteration (Rating: 1).

Time to Implement: Due to the high-level construction work for the shelving system, it may take several months to implement this solution in warehouses (Rating: 1).

3.5   Guided Examples of Evaluation Criteria

This section uses two independent mobile robot system assessment approaches using five previously mentioned multiple-decision criteria: the 'Equal Weight' approach and the Full Consistency Method (FUCOM). The 'Equal Weight' approach uses a scenario and the evaluation criteria for targeted decision-making. This approach is considered for its ease of use and establishing a benchmark for other decision methods. Then, as an alternative and recent multi-criteria decision-making method, FUCOM (Pamučar et al. 2018) is applied to the same set of criteria based on five supply chain experts' judgements.

3.5.1   The 'Equal Weight' Approach

In the first demonstration approach, all five criteria were assumed to have equal weight in the decision-making process. The rationale is further explained with a realistic scenario with a step-by-step decision tree to provide a guided example of the approach. The approach towards the scenario is not limited to mobile robot systems mentioned in this book but can include any others as it is based on the evaluation criteria rather than being system-specific (Fig. 3.7).

Mobile Robot Selection Scenario: A Logistics Service Provider in E-Commerce Fulfilment
The company is a logistics service provider in the e-commerce fulfilment market and serves several large e-commerce retailers. The company's upper management is confident that e-commerce will continue to grow due to customers' tendency to buy online, which has further increased during the Covid-19 pandemic. Hence, they want to increase the throughput of their warehouse and are seeking a scalable automation solution. The current portfolio includes small- to medium-sized products handled through manual case picking and broken-case picking. The company is not looking to expand their product portfolio through bulky products. The company might expand its operational area by purchasing the adjacent warehousing space in the logistics park if they see the targeted throughput growth. Thus, the upper management would like the system to adapt to potential layout alterations. Even though the upper management is motivated to implement a mobile robot system, considering the company's financial instability, it prefers an affordable solution. Finally, the management wants to implement the solution as soon as possible to recruit new customers.

3.5.2   FUCOM Approach

In case decision problems are characterised by criteria with differing degrees of importance, criteria should be allowed to have varying weights to reflect the decision-maker's preferences while minimising subjectivity in the process. As an alternative and independent assessment of mobile robot systems based on expert evaluations, FUCOM determines values for the weight coefficients of all criteria by performing consistent comparisons

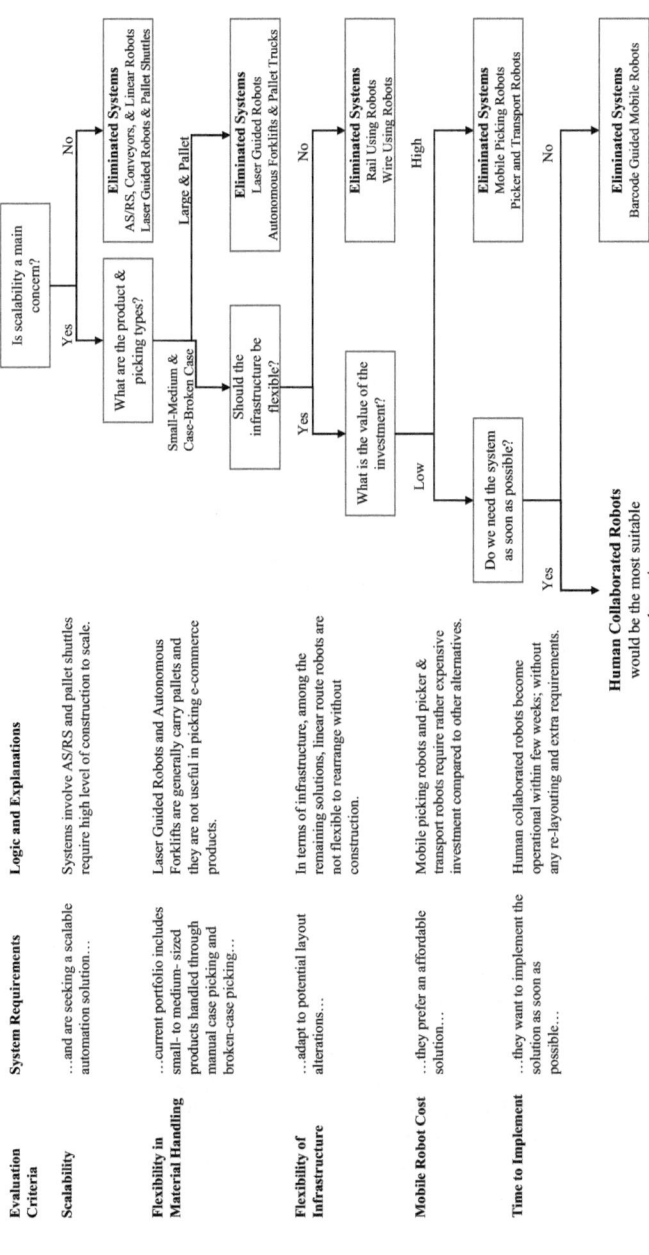

Fig. 3.7 Decision tree of the scenario with explanations. (Source: Authors)

based on supply chain experts' opinions. Pamučar et al. (2018) provide the foundation and details of the FUCOM.

As FUCOM uses expert evaluations, before FUCOM steps, the authors approached five supply chain experts to act as decision-makers in this study. All supply chain experts have been working at a large-sized company (headcount 250+) for more than ten years (Table 3.4).

Initially, the list of evaluation criteria is prepared in alphabetical order as C_1: Flexibility in infrastructure (max), C_2: Flexibility in handling (max), C_3: Mobile robot cost (min), C_4: Scalability (max), C_5: Time to implement (min).

Then, as the FUCOM process suggests, the predefined set of criteria is ranked by a supply chain expert (Decision-Maker 1) according to the significance of the criteria, that is, starting from the criterion which is expected to have the highest significance to the criterion of the least significance:

$$C_2 > C_1 > C_3 > C_5 > C_4$$

Once criteria are ranked, the decision-maker determines their importance coefficients with pairwise comparisons based on a scale [1–9], where '1' means equal importance and '9' means nine times more important compared to the next most important criterion (Pamučar et al. 2018). The pairwise priority comparison is always made with respect to the first-ranked C_2 criterion (Table 3.5).

Table 3.4 Information about experts

Experts	Years of experience	Role	Company sector	Company headcount
Expert 1	17	Strategic relationship manager (supply chain)	Aviation	680
Expert 2	20	Head of logistics engineering	Automotive	4160
Expert 3	27	Logistics transformation project manager	Food products and beverages	270,000
Expert 4	10	Senior lecturer in logistics and supply chain	University	3000
Expert 5	20	Head of accelerated digitalisation	Logistics provider	500,000

Source: Authors

Table 3.5 Priorities of criteria for decision-maker 1

Criterion	C_2	C_1	C_3	C_5	C_4
$\varpi_{C_{j(k)}}$	1	2	3	5	7

Source: Authors

Based on obtained priorities, comparative priorities are determined:

$$\phi_{C_2/C_1} = 2/1 = 2; \quad \phi_{C_1/C_3} = 3/2 = 1.5; \quad \phi_{C_3/C_5} = 5/3 = 1.\overline{6}; \quad \phi_{C_5/C_4} = 7/5 = 1.4$$

where 'k' means the rank of the criterion, the comparative priority ($\phi_{k/(k+1)}$) represents the significance or the priority that the criterion of $C_{j(k)}$ ranks over the criterion of $C_{j(k+1)}$ rank. For instance, if the criterion of $C_{j(k)}$ rank has the same significance as the criterion of $C_{j(k+1)}$ rank, then the comparative priority is $\phi_{k/(k+1)} = 1$.

In the third step, the final values of the weight coefficients of the evaluation criteria ($w_1, w_2, ..., w_n)^T$ are calculated. The final values of the weight coefficients should satisfy the following two conditions:

(a) the ratio of the weight coefficients is equal to the comparative priority among the observed criteria:

$$\frac{w_2}{w_1} = 2; \quad \frac{w_1}{w_3} = 1.5; \quad \frac{w_3}{w_5} = 1.\overline{6}; \quad \frac{w_5}{w_4} = 1.4$$

(b) in addition to the previous condition, the final values of the weight coefficients should meet the mathematical transitivity condition:

$$\frac{w_2}{w_3} = 2 * 1.5 = 3; \quad \frac{w_1}{w_5} = 1.5 * 1.\overline{6} = 2.4; \quad \frac{w_3}{w_4} = 1.\overline{6} * 1.4 = 2.\overline{3}$$

Our final model to determine the weight coefficients of the Decision-Maker 1 is as follows:

$$\min \chi$$

s.t.

$$\left|\frac{w_2}{w_1} - 2\right| \le \chi, \quad \left|\frac{w_1}{w_3} - 1.5\right| \le \chi, \quad \left|\frac{w_3}{w_5} - 1.\overline{6}\right| \le \chi, \quad \left|\frac{w_5}{w_4} - 1.4\right| \le \chi,$$

$$\left|\frac{w_2}{w_3} - 2*1.5\right| \le \chi, \quad \left|\frac{w_1}{w_5} - 1.5*1.\overline{6}\right| \le \chi, \quad \left|\frac{w_3}{w_4} - 1.\overline{6}*1.4\right| \le \chi,$$

$$\sum_{j=1}^{5} w_j = 1, \forall_j$$

$$w_j \ge 0, \forall_j$$

By solving this non-linear model, we obtain the final values of weight coefficients for flexibility in infrastructure, flexibility in handling, mobile robot cost, scalability, and time to implement as $(0.23, 0.46, 0.15, 0.07, 0.09)^{\mathrm{T}}$ and the deviation from a full consistency, $\chi = 0.00$ which means the results are reliable (Pamučar et al. 2018).

These steps are repeated with the other supply chain experts (Decision-Makers 2, 3, 4, 5) (Table 3.6). We assume the experts have comparable expertise and knowledge of the warehouse operations; hence, we use the average of individual weights to find the final weights of the criteria. The average resulting weights from five decision-makers are Flexibility in infrastructure: 0.16; Flexibility in material handling: 0.2; Mobile robot cost: 0.3; Scalability: 0.2; and Time to implement: 0.14.

As we multiply the average resulting weights of these criteria with our ratings for the mobile robot systems (Table 3.3) and sum the scores obtained from each criterion, a ranking of the mobile robot systems is produced. After the calculation, linear route mobile robots appear to be the best alternative for a generic warehouse, followed by human-collaborated robots and barcode-guided robots (Table 3.7). The two runner-up systems in FUCOM (human-collaborated robots and barcode-guided robots) were the final two candidates in the benchmark method, the 'Equal Weight' approach, presented in Fig. 3.7.

Table 3.6 Decision-makers and the FUCOM details

	Decision-maker 1	Decision-maker 2	Decision-maker 3	Decision-maker 4	Decision-maker 5
Criterion ranking	$C_2 > C_1 > C_3 > C_5 > C_4$	$C_3 > C_5 = C_1 = C_2 = C_4$	$C_4 > C_1 > C_3 > C_2 = C_5$	$C_3 > C_4 > C_2 > C_1 > C_5$	$C_3 > C_2 = C_4 = C_5 > C_1$
Priorities of criteria	1-2-3-5-7	1-2-4-4-4	1-2-3-5-5	1-2-3-3.5-4	1-2-2-2-3
Weight coefficients	0.23-0.46-0.15-0.07-0.09	0.44-0.22-0.11-0.11-0.11	0.45-0.22-0.15-0.09-0.09	0.42-0.21-0.14-0.12-0.10	0.35-0.18-0.18-0.18-0.12
Deviation from consistency	0.00	0.00	0.00	0.00	0.00
Average weight coefficients	$C_1-C_2-C_3-C_4-C_5$ 0.16-0.20-0.30-0.20-0.14				

Source: Authors

Table 3.7 Mobile robot systems and their ranking with weighted criteria

Mobile robot systems	Results
Rail- or wire-using robots	**2.34**
Human-collaborated robots	2.30
Barcode-guided robots	2.20
Autonomous forklifts	2.00
Mobile picking robots	2.00
Picker and transport robots	2.00
Laser-guided robots	1.80
AS/RS, conveyors, linear route robots	1.61
Laser-guided robots and pallet shuttles	1.61

Source: Authors

3.6   FINAL REMARKS

Such endeavours are complex decision problems with multiple dimensions, and implementing a solution affects the future capabilities of the warehouse. Thus, along with the identification of the correct system, decisions need to be made on three different levels to implement a system: strategic, tactical, and operational, in order of precedence (Bechtsis et al. 2017; Fragapane et al. 2021; Le-Anh and De Koster 2006). These decisions are derived from the synthesis of 130 papers in this review (Appendix). In the Appendix, the 'not mentioned' term is used if the information is missing, is ambiguous, or is not detailed. Also, terminology inconsistencies are corrected throughout this table. For instance, in the 'robot type' column, some papers clearly mean AMRs even though they refer to 'AGVs'.

REFERENCES

Abbas, A. S., Mohamed, T. A., & Hazem, M. (2018). *Optimization of Warehouse Material Handling Parameters to Enhance the Efficiency of Automated Sorting and Storage Systems. 11*(1), 2018.
Azadeh, K., De Koster, R. B. M., & Roy, D. (2019). Robotized and automated warehouse systems: Review and recent developments. In *Transportation Science* (Vol. 53, Issue 4). https://doi.org/10.1287/trsc.2018.0873
Banker, S. (2018). *Laser Guided Vehicles Navigate With Precision.* https://www.forbes.com/sites/stevebanker/2018/08/29/laser-guided-vehicles-navigate-with-precision/?sh=41e5e5465091
Bauters, K., De Cock, K., Hollevoet, J., Dobbelaere, G., & Van Landeghem, H. (2016). A simulation model to compare autonomous vehicle based ware-

houses with traditional AS/RS systems. *European Simulation and Modelling Conference 2016, ESM 2016*, 451–460.

Bechtsis, D., Tsolakis, N., Vlachos, D., & Iakovou, E. (2017). Sustainable supply chain management in the digitalisation era: The impact of Automated Guided Vehicles. *Journal of Cleaner Production, 142*, 3970–3984. https://doi.org/10.1016/j.jclepro.2016.10.057

Benady, D. (2016). *Will a robot take my job?* DHL. https://www.dhl.com/global-en/home/about-us/delivered-magazine/articles/2016/issue-3-2016/will-a-robot-take-my-job.html

Bogue, R. (2016). Growth in e-commerce boosts innovation in the warehouse robot market. *Industrial Robot, 43*(6), 583–587. https://doi.org/10.1108/IR-07-2016-0194

Bormann, R., de Brito, B. F., Lindermayr, J., Omainska, M., & Patel, M. (2019). Towards Automated Order Picking Robots for Warehouses and Retail. *Lecture Notes in Computer Science (Including Subseries Lecture Notes in Artificial Intelligence and Lecture Notes in Bioinformatics), 11754 LNCS*, 185–198. https://doi.org/10.1007/978-3-030-34995-0_18

Boysen, N., De Koster, R. B. M., & Weidinger, F. (2019). Warehousing in the e-commerce era: A survey. In *European Journal of Operational Research* (Vol. 277, Issue 2, pp. 396–411). Elsevier B.V. https://doi.org/10.1016/j.ejor.2018.08.023

Cribley, D. (2014). *9 Steps to Determining if an ASRS is Right For You.*

De Koster, R. B. M. (2018). Automated and robotic warehouses: developments and research opportunities. *Logistics and Transport, 2*(38), 33–40. https://doi.org/10.26411/83-1734-2015-2-38-4-18

Dematic. (n.d.). *CASE STUDY Brampton, Ontario, Canada*. Retrieved February 13, 2022, from https://www.dematic.com/en/downloads-and-resources/case-studies/download/?map=10&id=137

Dexion. (n.d.). *Pallet Shuttle System*. Retrieved June 5, 2022, from https://www.dexion.com/campaigns/semi-automation/shuttle/

DHL. (n.d.-a). *Focus Sheet Autonomous Forklift Deployment*. Retrieved February 13, 2022, from https://dhlinsights.dhlsupplychain.dhl.com/autonomous-forklift/focus-sheet_autonomous-forklift-deployment

Draganjac, I., Miklic, D., Kovacic, Z., Vasiljevic, G., & Bogdan, S. (2016). Decentralized Control of Multi-AGV Systems in Autonomous Warehousing Applications. *IEEE Transactions on Automation Science and Engineering, 13*(4), 1433–1447. https://doi.org/10.1109/TASE.2016.2603781

Enright, J. J., & Wurman, P. R. (2011). Optimization and coordinated autonomy in mobile fulfillment systems. *AAAI Workshop - Technical Report, WS-11-09*, 33–38.

Expeditors. (n.d.). *Case Study Continuous Improvement & Innovation: Robotics Technology in the Modern Warehouse*. Retrieved February 13, 2022, from https://info.expeditors.com/case-study-landing-north-coast-medical-robotics

Fan, Z., Gu, C., Yin, X., Liu, C., & Huang, H. (2018). Time window based path planning of multi-AGVs in logistics center. *Proceedings - 2017 10th International Symposium on Computational Intelligence and Design, ISCID 2017, 2,* 161–166. https://doi.org/10.1109/ISCID.2017.40

Ferrara, A., Gebennini, E., & Grassi, A. (2014). Fleet sizing of laser guided vehicles and pallet shuttles in automated warehouses. *International Journal of Production Economics, 157*(1), 7–14. https://doi.org/10.1016/j.ijpe.2014.06.008

Fragapane, G., de Koster, R., Sgarbossa, F., & Strandhagen, J. O. (2021). Planning and control of autonomous mobile robots for intralogistics: Literature review and research agenda. *European Journal of Operational Research, 294*(2), 405–426. https://doi.org/10.1016/j.ejor.2021.01.019

Füßler, D., Boysen, N., & Stephan, K. (2019). Trolley line picking: storage assignment and order sequencing to increase picking performance. *OR Spectrum, 41*(4), 1087–1121. https://doi.org/10.1007/s00291-019-00566-9

Geek+. (n.d.-a). *Forklift.* Geek+. Retrieved February 21, 2021, from https://www.geekplus.com/robot/f-robot

Geek+. (n.d.-b). *Geek+.* Geek+. Retrieved February 21, 2021, from https://www.geekplus.com/

Geek+. (n.d.-c). *How did Sinopharm Logistics break through its efficiency bottleneck?* Retrieved February 13, 2022, from https://blog.geekplus.com/case-studies/how-did-sinopharm-logistics-break-through-its-efficiency-bottleneck

Geek+. (n.d.-d). *Suning Logistics smoothly operates intelligent warehouse, greatly improving picking efficiency and accuracy.* Retrieved February 13, 2022, from https://blog.geekplus.com/case-studies/suning-logistics-smoothly-operates-intelligent-warehouse-greatly-improving-picking-efficiency-and-accuracy

Hai Robotics. (2019). *HAIPICK Landed in SF DHL Apparel Warehouse.* https://www.hairobotics.com/en/news/newsDetails?id=2c989740799dcb2c0179c072be0f0032

Hanson, R., Medbo, L., & Johansson, M. I. (2018). Performance Characteristics of Robotic Mobile Fulfilment Systems in Order Picking Applications. *IFAC-PapersOnLine, 51*(11), 1493–1498. https://doi.org/10.1016/j.ifacol.2018.08.290

Hikrobotics. (n.d.-a). *DHL Express.* Retrieved February 13, 2022, from https://www.hikrobotics.com/en/mobilerobot/bestpractice/detail?id=6

Hikrobotics. (n.d.-b). *Yuto Packaging.* Retrieved February 13, 2022, from https://www.hikrobotics.com/en/mobilerobot/bestpractice/detail?id=12

Horňáková, N., Jurík, L., Hrablik Chovanová, H., Cagáňová, D., & Babčanová, D. (2019). AHP method application in selection of appropriate material handling equipment in selected industrial enterprise. *Wireless Networks*. https://doi.org/10.1007/s11276-019-02050-2

Huang, G. Q., Chen, M. Z. Q., & Pan, J. (2015). Robotics in ecommerce logistics. *HKIE Transactions Hong Kong Institution of Engineers*, *22*(2), 68–77. https://doi.org/10.1080/1023697X.2015.1043960

inVia Robotics. (n.d.). *Zero to automated in weeks not months*. InVia Robotics. Retrieved February 21, 2021, from https://www.inviarobotics.com/implementation/

Kattepur, A., Rath, H. K., Simha, A., & Mukherjee, A. (2018). Distributed optimization in multi-agent robotics for industry 4.0 warehouses. *Proceedings of the ACM Symposium on Applied Computing*, 808–815. https://doi.org/10.1145/3167132.3167221

Kimura, N., Ito, K., Fuji, T., Fujimoto, K., Esaki, K., Beniyama, F., & Moriya, T. (2015). Mobile dual-arm robot for automated order picking system in warehouse containing various kinds of products. *2015 IEEE/SICE International Symposium on System Integration (SII)*, 332–338. https://doi.org/10.1109/SII.2015.7404942

Le-Anh, T., & De Koster, R. B. M. (2006). A review of design and control of automated guided vehicle systems. *European Journal of Operational Research*, *171*(1), 1–23. https://doi.org/10.1016/j.ejor.2005.01.036

Lee, H. Y., & Murray, C. C. (2019). Robotics in order picking: evaluating warehouse layouts for pick, place, and transport vehicle routing systems. *International Journal of Production Research*, *57*(18), 5821–5841. https://doi.org/10.1080/00207543.2018.1552031

Liu, S. (2018). Research on scheduling policy of automated warehouse system. *ACM International Conference Proceeding Series*, 1–5. https://doi.org/10.1145/3207677.3277965

Liu, Y., Ji, S., Su, Z., & Guo, D. (2019). Multi-objective AGV scheduling in an automatic sorting system of an unmanned (intelligent) warehouse by using two adaptive genetic algorithms and a multi-adaptive genetic algorithm. *PLoS ONE*, *14*(12). https://doi.org/10.1371/journal.pone.0226161

Locus Robotics. (n.d.-a). *Case Study: DHL Supply Chain*.

Locus Robotics. (n.d.-b). *Locus*. Locus Robotics. Retrieved February 21, 2021, from https://locusrobotics.com/why-locus/deployment/

Ly, G. B. (2019). Storage Assignment Policy and Route Planning of AGVS in Warehouse Optimization. In *Proceedings of 2019 International Conference on System Science and Engineering, ICSSE 2019*.

Magazino. (n.d.). Toru. Magazino. Retrieved February 21, 2021, from https://www.magazino.eu/products/toru/?lang=en

MiR. (n.d.). *ICM*. Retrieved February 13, 2022, from https://www.mobile-industrial-robots.com/case-studies/mir1000-icm-denmark/

Pamučar, D., Stević, Ž., & Sremac, S. (2018). A New Model for Determining Weight Coefficients of Criteria in MCDM Models: Full Consistency Method (FUCOM). *Symmetry, 10*(9), 393. https://doi.org/10.3390/sym10090393

Petković, T., Puljiz, D., Marković, I., & Hein, B. (2019). Human intention estimation based on hidden Markov model motion validation for safe flexible robotized warehouses. *Robotics and Computer-Integrated Manufacturing, 57*, 182–196. https://doi.org/10.1016/j.rcim.2018.11.004

Polten, L., & Emde, S. (2020). Scheduling automated guided vehicles in very narrow aisle warehouses. *Omega (United Kingdom)*. https://doi.org/10.1016/j.omega.2020.102204

Prime Vision. (n.d.). *Robotic Sorting*. Retrieved February 13, 2022, from https://www.primevision.com/robotic-sorting/

Raineri, M., Perri, S., & Guarino Lo Bianco, C. (2019). Safety and efficiency management in LGV operated warehouses. *Robotics and Computer-Integrated Manufacturing, 57*, 73–85. https://doi.org/10.1016/j.rcim.2018.11.003

Robots. (n.d.). *Fetch*. Retrieved April 27, 2022, from https://robots.ieee.org/robots/fetch/

Roy, D., Nigam, S., De Koster, R. B. M., Adan, I., & Resing, J. (2019). Robot-storage zone assignment strategies in mobile fulfillment systems. *Transportation Research Part E: Logistics and Transportation Review, 122*, 119–142. https://doi.org/10.1016/j.tre.2018.11.005

Schmidt, T., & Schulze, F. (2009). Future approaches to meet complexity requirements in material handling systems. *FME Transactions, 37*(4), 159–166.

Schulze, L., & Zhao, L. (2007). Worldwide development and application of automated guided vehicle systems. In *Int. J. Services Operations and Informatics* (Vol. 2, Issue 2).

Singhal, A., Singh, H. V., Penumatsa, A., Bhatt, N., Ambwani, P., Kumar, S., & Sinha, R. (2018). An actor based architecture for multi-robot system with application to warehouse. *IoPARTS 2018 - Proceedings of the 2018 International Workshop on Internet of People, Assistive Robots and ThingS*, 13–18. https://doi.org/10.1145/3215525.3215530

SSI Schaefer. (n.d.-a). *AGV Weasel*. SSI Schaefer. Retrieved February 21, 2021, from https://www.ssi-schaefer.com/en-us/products/conveying-transport/automated-guided-vehicles/fahrerloses-transportsystem-weasel-53020

SSI Schaefer. (n.d.-b). *Weasel*. SSI Schaefer. https://www.ssi-schaefer.com/resource/blob/56444/83497a9083769c1946bcc30f9f706473/brochure-weasel-en-dam-download-en-2054%2D%2Ddata.pdf

Swisslog. (n.d.-a). *CASE STUDY AMAG BUCHS NEAR ZURICH, SWITZERLAND*. Retrieved February 13, 2022, from https://www.swisslog.com/-/media/swisslog/documents/intralogistics-automation/case-studies/casestudy_amag_switzerland_eng_web.pdf?rev=9876e912b12043d29f331063340264b2&hash=1CB761D46A490DF465B6470A299DF53A

Swisslog. (n.d.-b). *Gries Deco Company*. Retrieved February 13, 2022, from https://www.swisslog.com/en-us/case-studies-and-resources/case-studies/2016/07/gries-deco-company

Wang, H., Chen, W., & Wang, J. (2019). Heterogeneous multi-agent routing strategy for robot-and-picker-to-good order fulfillment system. *Advances in Intelligent Systems and Computing, 867,* 237–249. https://doi.org/10.1007/978-3-030-01370-7_19

Wior, I., Jerenz, S., & Fay, A. (2018). Automated transportation systems subject to interruptions in production and intralogistics - a survey and evaluation. *International Journal of Logistics Systems and Management, 30*(4), 421. https://doi.org/10.1504/IJLSM.2018.093581

Wurman, P. R., DAndrea, R., & Mountz, M. (2008). *Coordinating Hundreds of Cooperative, Autonomous Vehicles in Warehouses.*

Yoshitake, H., Kamoshida, R., & Nagashima, Y. (2019). New Automated Guided Vehicle System Using Real-Time Holonic Scheduling for Warehouse Picking. *IEEE Robotics and Automation Letters, 4*(2), 1045–1052. https://doi.org/10.1109/LRA.2019.2894001

Yuan, R., Wang, H., & Li, J. (2019). The Pod Assignment Model and Algorithm in Robotic Mobile Fulfillment Systems. In Proceedings - IEEE International Conference on Service Operations and Logistics, and Informatics 2019, SOLI 2019. https://doi.org/10.1109/SOLI48380.2019.8955103

Zou, B., Xu, X., Gong, Y. (Yale), & De Koster, R. B. M. (2018). Evaluating battery charging and swapping strategies in a robotic mobile fulfillment system. *European Journal of Operational Research, 267*(2), 733–753. https://doi.org/10.1016/j.ejor.2017.12.008

CHAPTER 4

Strategic Decisions in Mobile Robot Automation

Strategic-level decisions involve comprehensive, long-term business decisions such as desired warehouse performance metrics and type of mobile robot systems. These decisions are generally taken before the installation stage of mobile robot systems, and they are rather hard to alter during the implementation. Effects of these decisions occur in the long run, which might be years.

4.1 Identifying the Criteria for Mobile Robot System Evaluation

Choosing a mobile robot system should depend on criteria analysis once a business has decided which warehouse operation/s it will automate. This book defines six criteria in Chap. 3, to evaluate mobile robot systems: mobile robot cost, flexibility of the infrastructure, flexibility of material handling, scalability, time to implement, and robustness. For instance, if there is a budget limit, that could be a constraint restricting available system alternatives. Furthermore, if the company has a growth expectation, the decision-maker might need to choose a scalable system. If the company is in a sector with different-sized products, the decision-maker might want to choose a flexible system in terms of material handling. If the warehouse layout and environment change frequently, systems with flexible infrastructures would be preferable. If the company is in an agile sector,

A. Yildirim et al., *Mobile Robot Automation in Warehouses*, Palgrave Studies in Logistics and Supply Chain Management, https://doi.org/10.1007/978-3-031-12307-8_4

such as e-commerce, time to implement criterion will help the decision-maker choose a system. Finally, as mobile drive units are expected to fail with a 36% probability within three months even if they undergo periodic maintenance (Yan et al. 2018), the system's behaviour towards these breakdowns can quickly become a significant criterion in choosing a system.

There could also be different criteria in different contexts. For example, Boysen et al. (2019) mention systems should be capable of handling large assortments of small orders and tight order deadlines with varying demand according to days of a year in e-commerce.

Operation or warehouse-specific strengths and weaknesses could also help the decision-maker determine the company's correct system as a criterion. Some automation systems, such as barcode-guided mobile robot systems, require a dedicated area to operate. Some automation systems, such as autonomous forklifts, would hardly work with small and medium-sized products. On the other hand, mobile picking robots cannot operate in a palletised-products-only environment.

Finally, simulations should be carried out while choosing the automation system to explore further and test the solution (Tsolakis et al. 2019). The decision-maker should be confident that the selected solution matches the company's expectations in terms of the aforementioned criteria.

4.2   IDENTIFYING KEY PERFORMANCE INDICATORS

Key performance indicators (KPIs) are generally a group of ratios that would help decision-makers assess the performance of the warehouse through tangible objectives and measures (Bartholdi and Hackman 2019; Rushton et al. 2014).

Decision-makers should use correct KPIs to measure the new system's success (Moeller et al. 2016). Thus, this paper divides KPIs into warehouse-specific KPIs and mobile robot system-specific KPIs to evaluate the performances of warehouse and automation systems separately and effectively.

4.2.1   Warehouse-Specific KPIs

These are traditional warehouse performance indicators that can be sub-categorised as: financial (costs, return on investment, capital), customer-related (customer retention, on-time delivery, quality), process-related (productivity, inventory levels, order picking error, utilisation), and worker-related (labour turnover, training, safety) (Bartholdi and Hackman 2019; Frazelle 2016; Hanson et al. 2018; Llopis-Albert et al. 2019; Rushton et al. 2014).

4.2.2   Mobile Robot System-Specific KPIs

These indicators refer to the performance of the chosen system and should be identified to develop the system and keep track of the potential performance improvement areas. For instance, Bechtsis et al. (2017, 2018) mention sustainability indicators for mobile robot systems such as the utilisation of robots, gas emission metrics, and resource efficiency metrics. Bauters et al. (2016) and Wang et al. (2020) consider system throughput (i.e. orders picked/hour) as the most important KPI for barcode-guided mobile robot systems, whereas Hanson et al. (2018) add hit rate (successive picks from the same shelf by the same person before the shelf was shifted) and Boysen et al. (2017) add the number of shelf moves. Füßler et al. (2019) use the walking distance of the picker as a key performance indicator in a linear route mobile robot system. This KPI could also be significant for human-collaborated mobile robots as it is a P2G system too (Wang et al. 2019a).

A decision-maker should choose among the mentioned mobile robot system KPIs and others in Table 4.1 to monitor the performance and effectiveness of the system at the macro level.

4.3   Type of Mobile Robots
and Their Coordination

As discussed in the previous chapters, two types of mobile robots are AGVs and AMRs. If the chosen system requires extensive onboard intelligence, such as the ability of dynamic path planning according to the environment and the obstacles, AMRs should be preferred. Otherwise, AGVs would be an acceptable and cheaper solution. It should be noted that AGVs only have sensors to stop when they face an obstacle and minimal onboard intelligence, which is not enough to calculate their path but is enough to get directions from a central mechanism.

The type of robots could be decided using various methods involving multiple criteria according to the necessities of the warehouse operations. For instance, Zavadskas et al. (2018) define seven criteria (investment,

Table 4.1   Mobile robot system KPIs

KPIs	Explanation	Considered by
System throughput	Orders or order lines picked per hour	Azadeh et al. (2019), Bauters et al. (2016), Draganjac et al. (2020), Fan et al. (2018), Lamballais et al. (2017, 2020), Lienert et al. (2019), Roy et al. (2019), Tai et al. (2018, 2019), Wang et al. (2019a, 2020), Wior et al. (2018), Yuan and Gong (2017), Zou et al. (2017, 2018)
Robot travel distance	Single robot travel distance on average in a time interval or cumulative distances	D'Emidio and Khan (2019), Feng et al. (2019), Haiming et al. (2019), Krnjak et al. (2015), Le-Anh and De Koster (2006), Lee and Murray (2019), Li et al. (2019), Merschformann et al. (2019), Panda et al. (2018), Qi et al. (2018), Santos et al. (2016), Vivaldini et al. (2016), Wang et al. (2020), Wei and Ni (2018), Weidinger et al. (2018), Yan et al. (2017), Yuan et al. (2019)
Robot utilisation/ idleness	Robot allocation or idleness percentages	Abbas et al. (2018), Azadeh et al. (2019), Bauters et al. (2016), De Koster et al. (2004), Dou et al. (2015), Draganjac et al. (2020), Ferrara et al. (2014), Ghassemi and Chowdhury (2018), Lamballais et al. (2017, 2020), Le-Anh et al. (2010), Llopis-Albert et al. (2019), Merschformann et al. (2019), Qi et al. (2018), Wang et al. (2019a), Wior et al. (2018), Yuan and Gong (2017)
Robot travel time	The cumulative time of robot travelling	Digani et al. (2019), Dou et al. (2015), Haiming et al. (2019), Le-Anh and De Koster (2006), Li et al. (2019), Liu (2018), Llopis-Albert et al. (2019), Panda et al. (2018), Qi et al. (2018), Wang et al. (2019a), Yan et al. (2017)
Avg. task completion time/ deadline miss	The average time required to fulfil a task	Confessore et al. (2013), D'Emidio and Khan (2019), Digani et al. (2019), Farinelli et al. (2017), Li et al. (2020), Semwal et al. (2018), Singhal et al. (2018), Vivaldini et al. (2016), Wior et al. (2018), Xue and Dong (2018), Yan et al. (2017), Zhang et al. (2018)
Task delay/ response time	The time it takes to find a robot for a task	Azadeh et al. (2019), De Koster et al. (2004), Le-Anh et al. (2010), Le-Anh and De Koster (2006), Li et al. (2019), Wior et al. (2018), Yan et al. (2017)

(*continued*)

Table 4.1 (continued)

KPIs	Explanation	Considered by
Computation/ negotiation time	The time it takes to calculate/negotiate a solution for a specific task group	Digani et al. (2016), Draganjac et al. (2020), Farinelli et al. (2017), Ghassemi and Chowdhury (2018), Lee and Murray (2019), Li et al. (2020), Sarkar et al. (2018), Tai et al. (2019), Weidinger et al. (2018)
# of deadlocks/ conflicts/routing failures	Deadlocks, conflicts, and routing failures encountered in a time interval	Fan et al. (2018), Haiming et al. (2019), Hanson et al. (2018), Krnjak et al. (2015), Le-Anh and De Koster (2006), Qi et al. (2018), Wior et al. (2018), Yan et al. (2017)
Workstation utilisation / idleness	For instance, the utilisation of picking workstations	Bauters et al. (2016), Merschformann et al. (2019), Wior et al. (2018), Yoshitake et al. (2019), Yuan and Gong (2017), Zou et al. (2017, 2018)
Task density	In a zone or other specific area	Qi et al. (2018), Vivaldini et al. (2016)
Robot-human ratio	The number of human workers required to operate with a certain number of robots or vice versa	Yuan and Gong (2017)
Number of active robots	The number of robots required in the field	Fan et al. (2018)
Walking distance	Human pickers travel distance on average in a time interval	Füßler et al. (2019)
Hit rate	Successive picks from the same shelf by the same person	Hanson et al. (2018)
Deadline miss	Minimising the cumulative time of missing deadlines	Sarkar and Agarwal (2019)

Source: Authors

minimum and maximum lift heights, dimensions, robot capacity and speed, and battery capacities of robots) and use a rough range of value-based process method to evaluate mobile robot types. On the other hand, Horňáková et al. (2019) define different criteria (investment, flexibility in material handling, return on investment, number of human workers, maintenance intensity, safety, and technical/implementation intensity) to choose the type of mobile robot and use analytical hierarchy process methods for evaluation.

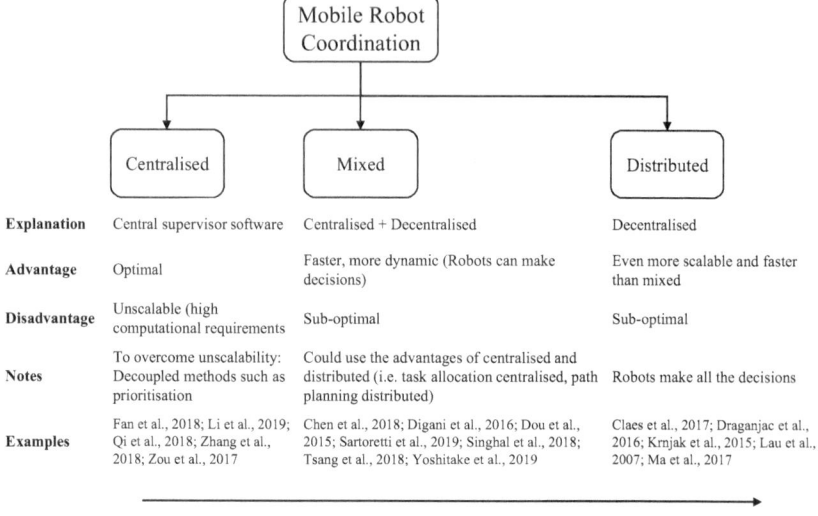

Fig. 4.1 Mobile robot coordination. (Source: Authors)

After choosing the type of mobile robots, their method of coordination should be determined for mobile robots' actions. Three main types of coordination methods are mentioned in the literature: centralised, distributed, and mixed (Fig. 4.1).

4.3.1   Centralised Coordination

In centralised systems, mobile robots are controlled by a central decision-making software which gathers all the information from the robots and the warehouse management software. Having gathered relevant operational information in a single source, a theoretically optimal solution is guaranteed (Tai et al. 2018). Furthermore, centralised coordination reduces the communication overhead of mobile robots and diminishes the delay which is caused by the deadlocks or conflicts of mobile robots (Li et al. 2019). Examples of centralised coordination can be seen in Li et al. (2019), Fan et al. (2018), and Qi et al. (2018). On the other hand, these systems are comparably unscalable with high computational requirements as the number of vehicles increases (Fan et al. 2018) and they have a single point failure (a potential error in the centralised mechanism disrupts the entire system) due to their centralised structure (Claes et al. 2017; Ghassemi and Chowdhury 2018; Singhal et al. 2018). Draganjac et al. (2020) claim that these systems can only control 10–15 robots.

4.3.2   Distributed Coordination

Accelerated artificial intelligence advancements encouraged a permanent solution to the scalability problem of centralised systems (Ghassemi and Chowdhury 2018). Distributed systems are more scalable, faster, and more dynamic to local information because intelligent mobile robots are given the authority to make decisions (Wen et al. 2018). However, mobile robots need to have a high level of autonomy to make these decisions (Krnjak et al. 2015). Thus, AGVs cannot be used in distributed systems. Distributed systems cannot reach global optimality (Fan et al. 2018) or a solution at all (Draganjac et al. 2016) since they do not have all the information in a single source. Ghassemi and Chowdhury (2018) mention that decentralised systems stay within 7–28% of the optimality of a centralised solution while being faster in achieving a feasible path by 1–3 orders of magnitude. Examples can be listed as Claes et al. (2017), Draganjac et al. (2016), Krnjak et al. (2015), and Lau et al. (2007).

4.3.3   Mixed Coordination

Mixed systems could be composed of centralised and distributed systems to take advantage of global information, such as auction-based systems. They are faster than centralised coordination and have a higher potential to find a solution closer to the optimal when compared to distributed coordination (Singhal et al. 2018). Furthermore, mixing centralised and decentralised coordination is also used in operational task division. For instance, centralised coordination could be used in task allocation, while a distributed coordination could be utilised for path planning (Chen et al. 2018; Dou et al. 2015; Tsang et al. 2018; Yoshitake et al. 2019), or distributed coordination could be implemented in robot charging, while centralised coordination could be preferred in task allocation (Xu et al. 2019). Rather than using different coordination methods for different operational tasks, the method of coordination can even differ in the same operational task. For example, Digani et al. (2016) apply centralised coordination for macro-management of path planning, but they prefer distributed coordination for path planning in local zones.

Higher volumes and an increased variety of products handled in warehouses necessitate intelligent robots deployed in picking and packing. For this reason, the presence of AMRs is becoming more popular in practice. Moreover, centralised algorithms require extensive computational power to provide optimal solutions in chaotic warehouses. Instead, there is a tendency towards suboptimal distributed solutions rather than optimal centralised solutions.

4.4   FACILITY LAYOUT

Most mobile robot systems require a significant change in the layout once they are deployed in manually operated warehouses. Besides 'U' and 'I' warehouse layout alternatives evaluated in Le-Anh et al. (2010), decisions need to be made about the following subjects.

4.4.1   *Number of Floors*

Modern warehouses are generally 8.5 to 11 metres high (Bartholdi and Hackman 2019). However, many mobile robot systems, especially those that automate the picking of small products, do not require that height. For instance, Magazino's mobile picking robot TORU can reach up to 3 metres high and barcode-guided mobile robot systems have a shelf height of around 2 metres (Toru n.d.; Wulfraat 2012). Thus, one-floored warehouses now have the potential to double their storage area and throughput by installing a second floor, carrying the same operation there. Lienert et al. (2019) and Ma et al. (2014) studied multi-floor warehouse design with elevators and mobile robot automation.

4.4.2   *Number, Size, and Location of Pick-Up and Delivery Points*

Most research on warehouse layout subjects related to mobile robot automation focuses on the aisle, cross-aisle, shelf, picking workstation and replenishment workstation quantities and sizes, and where to locate them if necessary. However, papers are mainly concentrated on barcode-guided mobile robots, which is influenced by a research focus on Kiva robots (Lamballais et al. 2020; Weidinger et al. 2018).

Queueing network models investigating the movement of tasks, such as from a shelf to a picking workstation, help design various layouts. For instance, Wang et al. (2020) propose a layout design framework using a queueing network. Lamballais et al. (2017) estimate the performance of different warehouse layouts and the number of workstations and shelves in terms of order throughput, whereas Lamballais et al. (2020) investigate the proportion between the picking and replenishment workstations.

The other layout designing option is to model the warehouse layout with variables and constraints and apply algorithms or heuristics to optimise the throughput without queueing models. Kumar and Kumar (2018) and Qi et al. (2018) have empirical studies on various warehouse layouts, including vertical, horizontal, and fishbone layouts. Weidinger et al. (2018) and Yuan et al. (2019) study optimum shelf locations in a fixed layout. Feng et al. (2019) and Lee and Murray (2019) provide information on picking workstation locations in traditional and flying-v layouts. Polten and Emde (2020) work on very narrow aisles and conclude that it is a managerial decision which may require task scheduling. D'Emidio and Khan (2019)compare a different number of aisles and lengths in a warehouse. Finally, Wang et al. (2019b) model and compare single-deep and double-deep shelf layouts in a mobile robot system in various scenarios and propose optimal lane depth for those scenarios.

4.4.3   Flow Path Layout of Mobile Robots

Vis (2006) and Le-Anh and De Koster (2006) provide an overview of the warehouse layout design in the presence of mobile robots. They mention three flow paths as single loop, tandem, and conventional, with single or multiple parallel lane formations and unidirectional or bidirectional routes.

Mobile robots travel in a unidirectional loop with a fixed sequence in the single-loop flow path. However, they need to pass through each location once before returning to the desired location the second time. Moreover, if a mobile robot stops processing a loading or an unloading operation, the vehicles behind it also pause (Le-Anh and De Koster 2006; Qi et al. 2018).

Tandem flow path overcomes these problems partially by installing multiple (parallel and/or perpendicular with connection points) single-loop flow paths. However, these systems are still not scalable for chaotic environments with many types of products and customer orders (Vis 2006).

Conventional flow path is a complex and scalable solution with multiple lanes and hybrid bidirectional and unidirectional routes. This type of layout requires a detailed and thorough traffic analysis to manage conflicts and deadlocks in a complex set of routes (Qi et al. 2018).

4.4.4   Idle Vehicle and Robot Charging Locations

While mobile robots are not assigned a task, they become idle and require a parking area (Vis 2006). A parking area should minimise the reaction time of mobile robots to new assignments and avoid potential congestion (Le-Anh and De Koster 2006). According to Le-Anh and De Koster (2006), idle vehicle parking locations are understudied, making empirical works or simulations unrealistic.

Another subject that is generally overlooked or omitted is robot charging locations. Many studies consider charging stations out of the operational area to simplify the traffic. However, some studies (Hamann et al. 2018; Le-Anh and De Koster 2006; Zou et al. 2018) question this decision and consider robot charging locations in the operational area to guarantee a close charging location.

4.5   Managing Human-Robot Interactions

When it comes to warehouse automation using mobile robots, the interaction between humans and robots and the role of human workers in deploying this disruptive technology in warehouses receive little attention, with scarce research on these subjects (Azadeh et al. 2019; Boysen et al. 2019). However, many automation systems involve human-robot interaction, as the two agents operating in either the same system or the same environment. For this reason, several subjects related to human-robot interaction should be considered before applying the chosen mobile robot system. If necessary, a change management team should be formed before introducing an automation system in the warehouse (Moeller et al. 2016) to address the concerns of existing employees and answer questions on how the operations are expected to change in the presence of a robot workforce.

4.5.1   Human Tasks

Even though automation technologies are deeply investigated, there are still several tasks, mainly related to grasping, that are carried out more efficiently by human workers than robots. Other examples of tasks where humans perform better than robots are maintenance of robots, picking, broken-case replenishment, and packing tasks. Thus, human-robot interaction is still inevitable, and the distribution of tasks should be done according to both sides' capabilities to maximise the joint performance and minimise errors (Azadeh et al. 2019; Bechtsis et al. 2017). Furthermore, distributed tasks should be optimised jointly as well. For instance, in human-collaborated mobile robot systems, if the human workers are standing in an aisle and mobile robots are visiting them one by one, task sequencing and path planning for both types of workers should be considered together as one optimisation task (Boysen et al. 2019).

4.5.2   Ergonomics

Ensuring the minimisation of the physical discomfort of human workers while distributing tasks between human workers and robots is another subject that needs consideration. The workforce in many industrialised countries has an ageing population, and workstations and tasks designed for human workers should take ergonomics into account (Boysen et al. 2019). Otherwise, human worker efficiency would be reduced because of excessive physical stress and the risk of injuries (Bechtsis et al. 2017). Furthermore, the insights of the interviews of Hanson et al. (2018) suggest that it is easier to recruit human workers as the ergonomics and working conditions improve.

Some mobile robot systems need workstations, and some mobile robot systems need a human worker to accompany them. Thus, the necessities of the systems should be identified, and the physical comfort of human workers should be planned and maximised beforehand.

4.5.3   Personnel Management

With the introduction of an automation solution, human workers feel a risk of job loss since some of the tasks are delegated to robots (Bechtsis et al. 2017; Moeller et al. 2016). For this reason, once works are distributed and optimised in automated warehouses, human workers should be trained and incentivised to increase the efficiency and throughput of the system (Azadeh et al. 2019; Bechtsis et al. 2017; Boysen et al. 2019). Moeller et al. (2016) discuss the effects of an automation system implementation through a real-life example as a case study and propose ways of communicating with human workers (events, meetings, posters, personalised merchandise about the solution, ensuring the focus is still on human workers) for successful change management application.

4.5.4   Human Safety

The introduction of mobile robots to the warehouses improved human safety due to unpersonned operation possibilities. For example, human-operated forklifts cause an accident every three days on average (Bechtsis et al. 2017). However, especially with the introduction of freely moving mobile robots, human safety has threatened and necessitated managerial attention (Inam and Raizer 2018; Raineri et al. 2019). Robot-human collaboration is inevitable, and there are systems with human-collaborative mobile robots in which human workers and robots need to operate in the same environment without boundaries and in synchronisation.

A risk assessment study is required to identify unsafe scenarios in the simulation stage of a mobile robot system implementation (Inam and Raizer 2018). Further, mobile robot multi-sensory systems could be used to increase vehicle awareness of the surrounding environment and decrease potential accidents (Bechtsis et al. 2017). Velocity planners (Raineri et al. 2019) and reinforcement learning applications (Sartoretti et al. 2019) are used to devise efficient trajectories while ensuring human safety. Petković et al. (2019) apply a Bayesian theory of mind approach for robots to estimate humans' intentions and avoid entering areas which humans will occupy. Lastly, human vision could be enhanced through augmented reality showing workers the planned path of mobile robots and allowing them to draw virtual walls to prevent mobile robots from entering those safe areas (Papcun et al. 2019).

REFERENCES

Abbas, A. S., Mohamed, T. A., & Hazem, M. (2018). Optimization of Warehouse Material Handling Parameters to Enhance the Efficiency of Automated Sorting and Storage Systems. 11(1), 2018.
Azadeh, K., De Koster, R. B. M., & Roy, D. (2019). Robotized and automated warehouse systems: Review and recent developments. In Transportation Science (Vol. 53, Issue 4). https://doi.org/10.1287/trsc.2018.0873
Bartholdi, J. J., & Hackman, S. T. (2019). Warehouse & Distribution Science. In Available on line at: http://www.tli.gatech.edu/ ... (Issue August).

Bauters, K., De Cock, K., Hollevoet, J., Dobbelaere, G., & Van Landeghem, H. (2016). A simulation model to compare autonomous vehicle based warehouses with traditional AS/RS systems. European Simulation and Modelling Conference 2016, ESM 2016, 451–460.

Bechtsis, D., Tsolakis, N., Vlachos, D., & Iakovou, E. (2017). Sustainable supply chain management in the digitalisation era: The impact of Automated Guided Vehicles. Journal of Cleaner Production, 142, 3970–3984. https://doi.org/10.1016/j.jclepro.2016.10.057

Bechtsis, D., Tsolakis, N., Vlachos, D., & Srai, J. S. (2018). Intelligent Autonomous Vehicles in digital supply chains: A framework for integrating innovations towards sustainable value networks. Journal of Cleaner Production, 181, 60–71. https://doi.org/10.1016/j.jclepro.2018.01.173

Boysen, N., Briskorn, D., & Emde, S. (2017). Parts-to-picker based order processing in a rack-moving mobile robots environment. European Journal of Operational Research, 262(2), 550–562. https://doi.org/10.1016/j.ejor.2017.03.053

Boysen, N., De Koster, R. B. M., & Weidinger, F. (2019). Warehousing in the e-commerce era: A survey. In European Journal of Operational Research (Vol. 277, Issue 2, pp. 396–411). Elsevier B.V. https://doi.org/10.1016/j.ejor.2018.08.023

Chen, H., Wang, Q., Yu, M., Cao, J., & Sun, J. (2018). Path Planning for Multi-robot Systems in Intelligent Warehouse. Lecture Notes in Computer Science (Including Subseries Lecture Notes in Artificial Intelligence and Lecture Notes in Bioinformatics), 11226 LNCS, 148–159. https://doi.org/10.1007/978-3-030-02738-4_13

Claes, D., Oliehoek, F., Baier, H., & Tuyls, K. (2017). Decentralised Online Planning for Multi-Robot Warehouse Commissioning.

Confessore, G., Fabiano, M., & Liotta, G. (2013). A network flow based heuristic approach for optimising AGV movements. Journal of Intelligent Manufacturing, 24(2), 405–419. https://doi.org/10.1007/s10845-011-0612-7

D'Emidio, M., & Khan, I. (2019). Collision-free allocation of temporally constrained tasks in multi-robot systems. Robotics and Autonomous Systems, 119, 151–172. https://doi.org/10.1016/j.robot.2019.07.002

De Koster, R. B. M., Le-Anh, T., & Van Der Meer, J. R. (2004). Testing and classifying vehicle dispatching rules in three real-world settings. Journal of Operations Management, 22(4 SPEC. ISS.), 369–386. https://doi.org/10.1016/j.jom.2004.05.006

Digani, V., Hsieh, M. A., Sabattini, L., & Secchi, C. (2019). Coordination of multiple AGVs: a quadratic optimization method. Autonomous Robots, 43(3), 539–555. https://doi.org/10.1007/s10514-018-9730-9

Digani, V., Sabattini, L., & Secchi, C. (2016). A probabilistic Eulerian traffic model for the coordination of multiple AGVs in automatic warehouses. IEEE

Robotics and Automation Letters, 1(1), 26–32. https://doi.org/10.1109/LRA.2015.2505646

Dou, J., Chen, C., & Yang, P. (2015). Genetic Scheduling and Reinforcement Learning in Multirobot Systems for Intelligent Warehouses. Mathematical Problems in Engineering, 2015. https://doi.org/10.1155/2015/597956

Draganjac, I., Miklic, D., Kovacic, Z., Vasiljevic, G., & Bogdan, S. (2016). Decentralized Control of Multi-AGV Systems in Autonomous Warehousing Applications. IEEE Transactions on Automation Science and Engineering, 13(4), 1433–1447. https://doi.org/10.1109/TASE.2016.2603781

Draganjac, I., Petrović, T., Miklić, D., Kovačić, Z., & Oršulić, J. (2020). Highly-scalable traffic management of autonomous industrial transportation systems. Robotics and Computer-Integrated Manufacturing, 63. https://doi.org/10.1016/j.rcim.2019.101915

Fan, Z., Gu, C., Yin, X., Liu, C., & Huang, H. (2018). Time window based path planning of multi-AGVs in logistics center. Proceedings - 2017 10th International Symposium on Computational Intelligence and Design, ISCID 2017, 2, 161–166. https://doi.org/10.1109/ISCID.2017.40

Farinelli, A., Boscolo, N., Zanotto, E., & Pagello, E. (2017). Advanced approaches for multi-robot coordination in logistic scenarios. Robotics and Autonomous Systems, 90, 34–44. https://doi.org/10.1016/j.robot.2016.08.010

Feng, L., Qi, M., Hua, S., & Zhou, Q. (2019). Picking Station Location in Traditional and Flying-V Aisle Warehouses for Robotic Mobile Fulfillment System. In IEEE International Conference on Industrial Engineering and Engineering Management (Vols. 2019-Decem). https://doi.org/10.1109/IEEM.2018.8607301

Ferrara, A., Gebennini, E., & Grassi, A. (2014). Fleet sizing of laser guided vehicles and pallet shuttles in automated warehouses. International Journal of Production Economics, 157(1), 7–14. https://doi.org/10.1016/j.ijpe.2014.06.008

Frazelle, E. H. (2016). World-Class Warehousing And Material Handling (Second Edi). McGraw-Hill Education.

Füßler, D., Boysen, N., & Stephan, K. (2019). Trolley line picking: storage assignment and order sequencing to increase picking performance. OR Spectrum, 41(4), 1087–1121. https://doi.org/10.1007/s00291-019-00566-9

Ghassemi, P., & Chowdhury, S. (2018). Decentralized task allocation in multi-robot systems via bipartite graph matching augmented with fuzzy clustering. In ASME.

Haiming, L., Weidong, L., Mei, Z., & An, C. (2019). Algorithm of path planning based on time window for multiple mobile robots in warehousing system. In Chinese Control Conference, CCC (Vols. 2019-July). https://doi.org/10.23919/ChiCC.2019.8866325

Hamann, H., Markarian, C., Auf Der Heide, F. M., & Wahby, M. (2018). Pick, Pack, & Survive: Charging Robots in a Modern Warehouse based on Online Connected Dominating Sets. Leibniz International Proceedings in Informatics, LIPIcs, 100, 221–2213. https://doi.org/10.4230/LIPIcs.FUN.2018.22

Hanson, R., Medbo, L., & Johansson, M. I. (2018). Performance Characteristics of Robotic Mobile Fulfilment Systems in Order Picking Applications. IFAC-PapersOnLine, 51(11), 1493–1498. https://doi.org/10.1016/j.ifacol.2018.08.290

Horňáková, N., Juřík, L., Hrablik Chovanová, H., Cagáňová, D., & Babčanová, D. (2019). AHP method application in selection of appropriate material handling equipment in selected industrial enterprise. Wireless Networks. https://doi.org/10.1007/s11276-019-02050-2

Inam, R., & Raizer, K. (2018). Risk Assessment for Human-Robot Collaboration in an automated warehouse scenario.

Krnjak, A., Draganjac, I., Bogdan, S., Petrovic, T., Miklic, D., & Kovacic, Z. (2015). Decentralized control of free ranging AGVs in warehouse environments. In Proceedings - IEEE International Conference on Robotics and Automation (Vols. 2015-June, Issue June). https://doi.org/10.1109/ICRA.2015.7139465

Kumar, N. V., & Kumar, C. S. (2018). Development of collision free path planning algorithm for warehouse mobile robot. Procedia Computer Science, 133, 456–463. https://doi.org/10.1016/j.procs.2018.07.056

Lamballais, T., Roy, D., & De Koster, R. B. M. (2017). Estimating performance in a Robotic Mobile Fulfillment System. European Journal of Operational Research, 256(3), 976–990. https://doi.org/10.1016/j.ejor.2016.06.063

Lamballais, T., Roy, D., & De Koster, R. B. M. (2020). Inventory allocation in robotic mobile fulfillment systems. IISE Transactions, 52(1), 1–17. https://doi.org/10.1080/24725854.2018.1560517

Lau, H. Y. K., Wong, V. W. K., & Lee, I. S. K. (2007). Immunity-based autonomous guided vehicles control. Applied Soft Computing Journal, 7(1), 41–57. https://doi.org/10.1016/j.asoc.2005.02.003

Le-Anh, T., & De Koster, R. B. M. (2006). A review of design and control of automated guided vehicle systems. European Journal of Operational Research, 171(1), 1–23. https://doi.org/10.1016/j.ejor.2005.01.036

Le-Anh, T., De Koster, R. B. M., & Yu, Y. (2010). Performance evaluation of dynamic scheduling approaches in vehicle-based internal transport systems. International Journal of Production Research, 48(24), 7219–7242. https://doi.org/10.1080/00207540903443279

Lee, H. Y., & Murray, C. C. (2019). Robotics in order picking: evaluating warehouse layouts for pick, place, and transport vehicle routing systems. International Journal of Production Research, 57(18), 5821–5841. https://doi.org/10.1080/00207543.2018.1552031

Li, M. P., Sankaran, P., Kuhl, M. E., Ptucha, R., Ganguly, A., & Kwasinski, A. (2019). Task Selection by Autonomous Mobile Robots in A Warehouse Using Deep Reinforcement Learning. In Proceedings - Winter Simulation Conference (Vols. 2019-Decem). https://doi.org/10.1109/WSC40007. 2019.9004792

Li, Z., Barenji, A. V., Jiang, J., Zhong, R. Y., & Xu, G. (2020). A mechanism for scheduling multi robot intelligent warehouse system face with dynamic demand. Journal of Intelligent Manufacturing, 31(2), 469–480. https://doi. org/10.1007/s10845-018-1459-y

Lienert, T., Stigler, L., & Fottner, J. (2019). Failure-Handling Strategies For Mobile Robots In Automated Warehouses. ECMS 2019 Proceedings Edited by Mauro Iacono, Francesco Palmieri, Marco Gribaudo, Massimo Ficco, 199–205. https://doi.org/10.7148/2019-0199

Liu, S. (2018). Research on scheduling policy of automated warehouse system. ACM International Conference Proceeding Series, 1–5. https://doi. org/10.1145/3207677.3277965

Llopis-Albert, C., Rubio, F., & Valero, F. (2019). Fuzzy-set qualitative comparative analysis applied to the design of a network flow of automated guided vehicles for improving business productivity. Journal of Business Research, 101, 737–742. https://doi.org/10.1016/j.jbusres.2018.12.076

Ma, Y., Wang, H., Xie, Y., & Guo, M. (2014). Path planning for multiple mobile robots under double-warehouse. Information Sciences, 278, 357–379. https://doi.org/10.1016/j.ins.2014.03.058

Merschformann, M., Lamballais, T., De Koster, R. B. M., & Suhl, L. (2019). Decision rules for robotic mobile fulfillment systems. Operations Research Perspectives, 6. https://doi.org/10.1016/j.orp.2019.100128

Moeller, K., Gabel, J., & Bertagnolli, F. (2016). FISCHER FIXING SYSTEMS: MOVING FORWARD WITH THE WORKFORCE - CHANGE COMMUNICATION AT THE GLOBAL DISTRIBUTION CENTER. Journal of Information Technology Education: Discussion Cases, 5(September 2017), 01. https://doi.org/10.28945/3457

Panda, M. R., Das, P. K., Dutta, S., & Pradhan, S. K. (2018). Optimal path planning for mobile robots using oppositional invasive weed optimization. Computational Intelligence, 34(4), 1072–1100. https://doi.org/https:// doi.org/10.1111/coin.12166

Papcun, P., Cabadaj, J., Kajati, E., Romero, D., Landryova, L., Vascak, J., & Zolotova, I. (2019). Augmented Reality for Humans-Robots Interaction in Dynamic Slotting "Chaotic Storage" Smart Warehouses. IFIP Advances in Information and Communication Technology, 566, 633–641. https://doi. org/10.1007/978-3-030-30000-5_77

Petković, T., Puljiz, D., Marković, I., & Hein, B. (2019). Human intention estimation based on hidden Markov model motion validation for safe flexible robotized warehouses. Robotics and Computer-Integrated Manufacturing, 57, 182–196. https://doi.org/https://doi.org/10.1016/j.rcim.2018.11.004

Polten, L., & Emde, S. (2020). Scheduling automated guided vehicles in very narrow aisle warehouses. Omega (United Kingdom). https://doi.org/10.1016/j.omega.2020.102204

Qi, M., Li, X., Yan, X., & Zhang, C. (2018). On the evaluation of AGVS-based warehouse operation performance. Simulation Modelling Practice and Theory, 87, 379–394. https://doi.org/10.1016/j.simpat.2018.07.015

Raineri, M., Perri, S., & Guarino Lo Bianco, C. (2019). Safety and efficiency management in LGV operated warehouses. Robotics and Computer-Integrated Manufacturing, 57, 73–85. https://doi.org/10.1016/j.rcim.2018.11.003

Roy, D., Nigam, S., De Koster, R. B. M., Adan, I., & Resing, J. (2019). Robot-storage zone assignment strategies in mobile fulfillment systems. Transportation Research Part E: Logistics and Transportation Review, 122, 119–142. https://doi.org/10.1016/j.tre.2018.11.005

Rushton, A., Croucher, P., & Baker, P. (2014). Handbook of logistics and distribution management.

Santos, J., Costa, P., Rocha, L., Vivaldini, K., Moreira, A. P., & Veiga, G. (2016). Validation of a time based routing algorithm using a realistic automatic warehouse scenario. Advances in Intelligent Systems and Computing, 418, 81–92. https://doi.org/10.1007/978-3-319-27149-1_7

Sarkar, C., & Agarwal, M. (2019). Cannot avoid penalty? Let's minimize. In Proceedings - IEEE International Conference on Robotics and Automation (Vols. 2019-May). https://doi.org/10.1109/ICRA.2019.8794338

Sarkar, C., Paul, H. S., & Pal, A. (2018). A Scalable Multi-Robot Task Allocation Algorithm. In Proceedings - IEEE International Conference on Robotics and Automation. https://doi.org/10.1109/ICRA.2018.8460886

Sartoretti, G., Kerr, J., Shi, Y., Wagner, G., Satish Kumar, T. K., Koenig, S., & Choset, H. (2019). PRIMAL: Pathfinding via Reinforcement and Imitation Multi-Agent Learning. In IEEE Robotics and Automation Letters (Vol. 4, Issue 3). https://doi.org/10.1109/LRA.2019.2903261

Semwal, T., Jha, S. S., & Nair, S. B. (2018). On Ordering Multi-Robot Task Executions within a Cyber Physical System. ACM Transactions on Autonomous and Adaptive Systems, 12(4), 1–27. https://doi.org/10.1145/3124677

Singhal, A., Singh, H. V., Penumatsa, A., Bhatt, N., Ambwani, P., Kumar, S., & Sinha, R. (2018). An actor based architecture for multi-robot system with application to warehouse. IoPARTS 2018 - Proceedings of the 2018 International Workshop on Internet of People, Assistive Robots and ThingS, 13–18. https://doi.org/10.1145/3215525.3215530

Tai, R., Wang, J., & Chen, W. (2019). A prioritized planning algorithm of trajectory coordination based on time windows for multiple AGVs with delay disturbance. Assembly Automation, 39(5), 753–768. https://doi.org/10.1108/AA-03-2019-0054

Tai, R., Wang, J., Tian, W., Chen, W., Wang, H., & Zhou, Y. (2018). A Time-Efficient Approach to Solve Conflicts and Deadlocks for Scheduling AGVs in Warehousing Applications. 2018 IEEE International Conference on Real-Time Computing and Robotics (RCAR), 166–171. https://doi.org/10.1109/RCAR.2018.8621773

Toru. (n.d.). Magazino. Retrieved February 21, 2021, from https://www.maga-zino.eu/products/toru/?lang=en

Tsang, K. F. E., Ni, Y., Wong, C. F. R., & Shi, L. (2018). A Novel Warehouse Multi-Robot Automation System with Semi-Complete and Computationally Efficient Path Planning and Adaptive Genetic Task Allocation Algorithms.

Tsolakis, N., Bechtsis, D., & Srai, J. S. (2019). Intelligent autonomous vehicles in digital supply chains: From conceptualisation, to simulation modelling, to real-world operations. Business Process Management Journal, 25(3), 414–437. https://doi.org/10.1108/BPMJ-11-2017-0330

Vis, I. F. A. (2006). Survey of research in the design and control of automated guided vehicle systems. European Journal of Operational Research, 170(3), 677–709. https://doi.org/10.1016/j.ejor.2004.09.020

Vivaldini, K., Rocha, L. F., Martarelli, N. J., Becker, M., & Moreira, A. P. (2016). Integrated tasks assignment and routing for the estimation of the optimal number of AGVS. International Journal of Advanced Manufacturing Technology, 82(1–4), 719–736. https://doi.org/10.1007/s00170-015-7343-4

Wang, H., Chen, W., & Wang, J. (2019a). Heterogeneous multi-agent routing strategy for robot-and-picker-to-good order fulfillment system. Advances in Intelligent Systems and Computing, 867, 237–249. https://doi.org/10.1007/978-3-030-01370-7_19

Wang, K., Yang, Y., & Li, R. (2019b). Travel time models for the rack-moving mobile robot system. International Journal of Production Research. https://doi.org/10.1080/00207543.2019.1652778

Wang, W., Wu, Y., Zheng, J., & Chi, C. (2020). A comprehensive framework for the design of modular robotic mobile fulfillment systems. IEEE Access, 8, 13259–13269. https://doi.org/10.1109/ACCESS.2020.2966403

Wei, C., & Ni, F. (2018). Tabu temporal difference learning for robot path planning in uncertain environments. Lecture Notes in Computer Science (Including Subseries Lecture Notes in Artificial Intelligence and Lecture Notes in Bioinformatics), 10965 LNAI, 123–134. https://doi.org/10.1007/978-3-319-96728-8_11

Weidinger, F., Boysen, N., & Briskorn, D. (2018). Storage assignment with rack-moving mobile robots in KIVA warehouses. Transportation Science, 52(6), 1479–1495. https://doi.org/10.1287/trsc.2018.0826

Wen, J., He, L., & Zhu, F. (2018). Swarm Robotics Control and Communications: Imminent Challenges for Next Generation Smart Logistics. IEEE Communications Magazine, 56(7), 102–107. https://doi.org/10.1109/MCOM.2018.1700544

Wior, I., Jerenz, S., & Fay, A. (2018). Automated transportation systems subject to interruptions in production and intralogistics - a survey and evaluation. International Journal of Logistics Systems and Management, 30(4), 421. https://doi.org/10.1504/IJLSM.2018.093581

Wulfraat, M. (2012). A Supply Chain Consultant Evaluation of Kiva Systems (Amazonrobotics). https://mwpvl.com/html/kiva_systems.html

Xu, J., Wang, J., & Chen, W. (2019). An efficient recharging task planning method for multi-robot autonomous recharging problem. In IEEE International Conference on Robotics and Biomimetics, ROBIO 2019. https://doi.org/10.1109/ROBIO49542.2019.8961439

Xue, F., & Dong, T. (2018). Research on the logistics robot task allocation method based on improved ant colony algorithm. In Int. J. Wireless and Mobile Computing (Vol. 14, Issue 4).

Yan, R. D., Dunnett, S. J., & Jackson, L. M. (2018). Optimising the maintenance strategy for a multi-AGV system using genetic algorithms. In Safety and Reliability – Safe Societies in a Changing World. CRC Press. https://doi.org/10.1201/9781351174664

Yan, X., Zhang, C., & Qi, M. (2017). Multi-AGVs Collision-Avoidance and Deadlock-Control for Item-to-Human Automated Warehouse. IEEE International Conference on Robotics and Biomimetics, ROBIO 2019.

Yoshitake, H., Kamoshida, R., & Nagashima, Y. (2019). New Automated Guided Vehicle System Using Real-Time Holonic Scheduling for Warehouse Picking. IEEE Robotics and Automation Letters, 4(2), 1045–1052. https://doi.org/10.1109/LRA.2019.2894001

Yuan, R., Wang, H., & Li, J. (2019). The Pod Assignment Model and Algorithm in Robotic Mobile Fulfillment Systems. In Proceedings - IEEE International Conference on Service Operations and Logistics, and Informatics 2019, SOLI 2019. https://doi.org/10.1109/SOLI48380.2019.8955103

Yuan, Z., & Gong, Y. Y. (2017). Bot-in-time delivery for robotic mobile fulfillment systems. IEEE Transactions on Engineering Management, 64(1), 83–93. https://doi.org/10.1109/TEM.2016.2634540

Zavadskas, E. K., Nunić, Z., Stjepanović, Ž., & Prentkovskis, O. (2018). Novel Rough Range of Value Method (R-ROV) for selecting automatically guided vehicles (AGVs). Studies in Informatics and Control, 27(4), 385–394. https://doi.org/10.24846/v27i4y201802

Zhang, Z., Guo, Q., Chen, J., & Yuan, P. (2018). Collision-Free Route Planning for Multiple AGVs in an Automated Warehouse Based on Collision Classification. IEEE Access, 6, 26022–26035. https://doi.org/10.1109/ACCESS.2018.2819199

Zou, B., Gong, Y. (Yale), Xu, X., & Yuan, Z. (2017). Assignment rules in robotic mobile fulfilment systems for online retailers. International Journal of Production Research, 55(20), 6175–6192. https://doi.org/10.108 0/00207543.2017.1331050

Zou, B., Xu, X., Gong, Y. (Yale), & De Koster, R. B. M. (2018). Evaluating battery charging and swapping strategies in a robotic mobile fulfillment system. European Journal of Operational Research, 267(2), 733–753. https://doi. org/10.1016/j.ejor.2017.12.008

CHAPTER 5

Tactical Decisions in Mobile Robot Automation

Based on strategic decisions, tactical decisions need to be made next which have a medium-term effect on warehouse operations. These decisions are comparably easier to adjust during the implementation and ongoing operation compared to strategic-level decisions. Still, they should not and generally cannot be changed in the short term.

5.1   STORAGE ASSIGNMENT PLAN

Also referred to as quantity and type of products in pick locations, these decisions are aligned with facility layout decisions at the strategic level. Once storage locations are decided, products should be distributed in a way which maximises product availability and accessibility. Product distribution is not only limited to the quantities as Stock Keeping Units (SKUs) can have a wide variety of shapes and sizes, which might necessitate the usage of different types of shelves in storage locations (Lamballais et al. 2020). Even though it was a type of decision that remained unchanged for several months for traditional warehouses, it can now be altered at short notice. The reason is the ability of mobile robots that could carry and change shelf positions (barcode-guided mobile robots) or change product locations overnight without human assistance (i.e. autonomous forklifts, self-picking mobile robots) (Füßler et al. 2019).

According to Weidinger et al. (2018), it performs best in the presence of mobile robots among five different storage policies, that is, random storage (also see Feng et al. 2019; Lamballais et al. 2017), closest open location storage, dedicated storage, full-turnover storage, and ABC storage. A known strategy for storage assignment among these five strategies is ABC storage, which divides products on hand into three zones based on their turnover rate (De Koster 2018). Even though ABC storage is the most common approach, better results could be achieved with hybrid approaches such as first dividing the warehouse into zones according to the ABC storage and then applying the closest open location storage within these zones (Ly 2019).

Other than these five storage policies, uniform storage policy (Lee and Murray 2019) or storing products according to their desired service levels (He et al. 2018) could be other approaches. Lee and Murray (2019) apply a uniform storage policy.

The most frequently applied method to determine the pick location and the number of products on shelves is simulated annealing (Füßler et al. 2019; Merschformann et al. 2019; Yuan et al. 2019). It is also common to distribute the same SKUs to more than one shelf (mixed-shelves storage) to increase product availability (Boysen et al. 2019). Indeed, the more scattered SKUs to different shelves, the fewer mobile robots will be required to deliver the same throughput as some studies suggest (Bauters et al. 2016; Boysen et al. 2017; Lamballais et al. 2020).

Being informed by these studies, the storage assignment plan should be based on the chosen mobile robot system. For instance, in barcode-guided mobile robot systems, shelves are portable, and their locations are subject to change each time they are carried to the picking workstation. For this reason, it is common to use a random open location shelf storage policy where shelves are equally likely to be stored in any of the available locations (Roy et al. 2019). Still, to choose among those approaches, other decisions (picking order assignment, picking shelf selection, replenishment decisions) should be considered as well (Merschformann et al. 2019).

Increasing product availability and accessibility improves warehouse efficiency. Related studies on storage assignment with robots are limited, and further research is needed to evaluate different storage policies for the various mobile robot systems.

5.2   Order Management Plan

The order management plan is the strategy which answers the question: 'How to process incoming customer orders?'. This book divides order management strategies in mobile robot automation into static (offline) and dynamic (online).

5.2.1   Static Order Management

In static order management, customer orders are collected by the system and then divided into batches or waves with a cut-off point (Boysen et al. 2019; Rushton et al. 2014). For batches, the cut-off point could be a certain quantity of orders, while for waves, it could be a time of the day (Rushton et al. 2014).

Order batching or waves of orders could result in a list of products or shelves. If, for instance, barcode-guided mobile robot systems are implemented, they would yield a sequence of shelves that contain the necessary amount of each product (Boysen et al. 2017). Otherwise, they would produce a list of products to be picked and sorted via a mobile robot or a human picker. Boysen et al. (2019) did a detailed review of batch and wave of orders and their impact on warehouse productivity metrics.

As order batching in chaotic warehouses is computationally complex (NP-hard) and time-consuming, heuristics might be used to obtain feasible but suboptimal solutions (Li et al. 2017). Azadeh et al. (2019) add that linear programming or mixed-integer programming could help to determine batching rules. Zou et al. (2019) used the seed algorithm to batch and manage orders, but they found out that only large scale of orders (i.e. four items per order and more than 50 orders) could be efficiently handled. Boysen et al. (2017) apply simulated annealing to batch orders and conclude that sequencing orders present better solutions against sequencing shelves to be visited. Li et al. (2017) use a three-stage hybrid heuristic algorithm for shelf sequencing, but they sequence shelves only for one batch of orders. Merschformann et al. (2019) batch replenishment orders through random batching and shelf-based batching and find out that shelf-based batching is slightly better than random batching. Sarkar et al. (2018), Sarkar and Agarwal (2019), and Xue and Dong (2018) are the papers that use waves of orders as a static order management strategy. However, they did not analyse or compare the performance of waves in their studies.

5.2.2   Dynamic Order Management

In this version, customer orders, which are placed through the warehouse management system, are directly added at the end of picklist or get prioritised if they are urgent (Boysen et al. 2019). This way, urgent orders could quickly be dealt with that would minimise total tardiness of customer orders and order fulfilment system would be agile. Even though many papers, such as Lamballais et al. (2017) and Tai et al. (2018), use dynamic order management, none of them presents a performance comparison with static order management strategies.

Storage assignment and order management strategies should be jointly and dynamically determined to increase the throughput (He et al. 2018). Distributing the workload fairly among mobile robots or workstations would ensure all partial orders are finished nearly simultaneously, which would decrease the sortation and packing challenges or robot conflicts and deadlocks (Boysen et al. 2019).

5.3   QUANTITY OF ROBOTS (FLEET SIZING)

The variable cost of mobile robot systems is based on the number of robots bought (Wang et al. 2020). For this reason, considering the demand variability, fleet sizing is a decision that should be made carefully, as underestimating the fleet size would cause a delay in the orders and overestimating it would cause too much traffic (Vivaldini et al. 2016). This decision can be altered in the medium run with additional robots or decreasing the number of active robots due to seasonal demand changes.

There are two types of robot fleet behaviours. A 'dedicated' robot fleet can divide and dedicate itself to workstations or human workers or a 'pooled' robot fleet can act together homogeneously to maximise the throughput (Yuan and Gong 2017). Deciding on that strategy first helps the decision-maker evaluate the fleet size more systematically. Roy et al. (2019) and Yuan and Gong (2017) studied pooled and dedicated mobile robots. Roy et al. (2019) suggest that using pooled robots instead of dedicated robots reduces the throughput time for order picking up to one-third. Considering robot congestion and operable area, Yuan and Gong (2017) provide an optimal ratio for human pickers and robots for both alternatives as design support.

Ferrara et al. (2014) optimise the number of mobile robots through a queueing network, while Lamballais et al. (2017) and Merschformann et al. (2019) optimise the number of mobile robots through workstation utilisation. Boysen et al. (2017) relate the optimum number of robots with the SKU diversity on shelves and calculate the fleet size accordingly, whereas Sarkar et al. (2018) apply nearest-neighbour-based clustering and routing algorithms to find the required number of robots. Polten and Emde (2020) finish their study with a shortcut for autonomous forklifts operating in narrow aisles and advice decision-makers to have fewer forklifts than the number of aisles. Finally, Le-Anh and De Koster (2006) review various fleet sizing methods to aid the decision-makers. Yet, they emphasise the importance of simulations as fleet size could differ according to the task allocation algorithm and traffic congestions in practice.

Empirical papers that mention fleet sizing generally concentrate on proving their algorithms on various fleet sizes, and most of them do not go beyond fleet comparisons on throughput performances. However, fleet size should be considered as a variable to optimise as an outcome of the algorithm as it is the biggest portion of the variable cost of mobile robot systems.

5.4   MAINTENANCE AND FAILURE HANDLING STRATEGIES

Unmanned mobile robots can work continuously yet have lower maintenance costs compared to manned vehicles (Bechtsis et al. 2017). Maintenance costs could be lowered further if tasks are distributed evenly since their active time would be reduced, and their fixed maintenance would be prolonged (Weidinger et al. 2018). Still, maintenance of mobile robots should be planned, including backup strategies for their downtime, as a robot's failure could disrupt the entire system (Wior et al. 2018).

Once each subsystem's yearly failure rates are calculated and used in a simulation to get the task reliability values, it is observed that mobile robots fail the most at 'travelling to storage area' and 'travelling to workstation' tasks (Yan et al. 2017).

Yan et al. (2018) optimise the maintenance strategy by using preventive and corrective maintenance via coloured Petri nets and genetic algorithms and estimate yearly maintenance costs. They find that corrective maintenance provides long-term high efficiency of the mobile robot systems even though it is more expensive. They also state that, within twelve months, a mobile robot will fail with a 98% probability after completing 3280 tasks even though they undergo periodic maintenance, stating the importance of corrective maintenance.

When a robot fails, four strategies may be followed: (1) ignore the broken robot, (2) pause the whole system, (3) restart the system and replan tasks, (4) reroute tasks from the broken robot to other robots (Lienert et al. 2019). Rerouting only the robots affected by the malfunctioning robot's tasks can maximise the throughput the most as it does not stop the whole system (Confessore et al. 2013; Draganjac et al. 2016; Kattepur et al. 2018; Lau et al. 2007; Liu et al. 2019a; Sartoretti et al. 2019). Another method to include robot failure in simulations, Witczak et al. (2020) apply a predictive fault-tolerant control algorithm and guarantee that all tasks would be completed irrespective of the faults of the robots.

5.5   ROBOT ENERGY MANAGEMENT

Even though one of the main advantages of mobile robots is working 24/7 through battery charging, energy management is one of the least attended subjects at tactical and operational levels (Le-Anh and De Koster 2006; Vis 2006). Besides, efficient energy management would affect the fleet size of mobile robots as well as the throughput of the warehouse by increasing the robot availability (Vis 2006).

Using a semi-open queueing network, Zou et al. (2018) compare plug-in charging, inductive charging, and battery swapping as alternative strategies for robot charging plans. They concluded inductive charging is the best alternative, and even though battery swapping outperforms plug-in charging, it is a more expensive strategy. The problem with the inductive charging strategy is that not all robots in the mobile robot system solutions can be inductively charged.

Robots can create a charging task request based on their battery levels (Kattepur et al. 2018), or their battery levels could be calculated in time windows or task sequences to create a charging task for them (Xu et al. 2019). According to Hamann et al. (2018), Amazon charges its Kiva robots for 5 minutes every 55 minutes. Many studies, on the other hand, charge mobile robots based on their energy levels, such as 20%, 40%, and 50% (Confessore et al. 2013; Lee et al. 2019; Lee and Murray 2019; Semwal et al. 2018; Yoshitake et al. 2019; Zhang et al. 2018). As an alternative approach, Liu et al. (2019b) prefer creating charging tasks according to the power consumed by active time and speed.

The technology behind the energy capacity of robots is continuously developing, and it either powers a more robust robot or increases the active times of robots. Considering this fact, an efficient charging type should be determined specific to the case, and following that decision, a charging level should be set to avoid energy-based failures during the operation. The level could be set through the suggestions of papers provided (20%, 40%, or 50%) or through the trial-and-error method. Further, using mobile charging stations and minimising the number of charging locations while covering the whole operational area could significantly increase the availability of mobile robots (Hamann et al. 2018). Finally, most of the studies are on batteries. However, electric-fuelled robots might provide a more sustainable and environment-friendly warehouse (Bechtsis et al. 2017).

REFERENCES

Azadeh, K., De Koster, R. B. M., & Roy, D. (2019). Robotized and automated warehouse systems: Review and recent developments. In Transportation Science (Vol. 53, Issue 4). https://doi.org/10.1287/trsc.2018.0873

Bauters, K., De Cock, K., Hollevoet, J., Dobbelaere, G., & Van Landeghem, H. (2016). A simulation model to compare autonomous vehicle based warehouses with traditional AS/RS systems. European Simulation and Modelling Conference 2016, ESM 2016, 451–460.

Bechtsis, D., Tsolakis, N., Vlachos, D., & Iakovou, E. (2017). Sustainable supply chain management in the digitalisation era: The impact of Automated Guided Vehicles. Journal of Cleaner Production, 142, 3970–3984. https://doi.org/10.1016/j.jclepro.2016.10.057

Boysen, N., Briskorn, D., & Emde, S. (2017). Parts-to-picker based order processing in a rack-moving mobile robots environment. European Journal of Operational Research, 262(2), 550–562. https://doi.org/10.1016/j.ejor.2017.03.053

Boysen, N., De Koster, R. B. M., & Weidinger, F. (2019). Warehousing in the e-commerce era: A survey. In European Journal of Operational Research (Vol. 277, Issue 2, pp. 396–411). Elsevier B.V. https://doi.org/10.1016/j.ejor.2018.08.023

Confessore, G., Fabiano, M., & Liotta, G. (2013). A network flow based heuristic approach for optimising AGV movements. Journal of Intelligent Manufacturing, 24(2), 405–419. https://doi.org/10.1007/s10845-011-0612-7

De Koster, R. B. M. (2018). Automated and robotic warehouses: developments and research opportunities. Logistics and Transport, 2(38), 33–40. https://doi.org/10.26411/83-1734-2015-2-38-4-18

Draganjac, I., Miklic, D., Kovacic, Z., Vasiljevic, G., & Bogdan, S. (2016). Decentralized Control of Multi-AGV Systems in Autonomous Warehousing Applications. IEEE Transactions on Automation Science and Engineering, 13(4), 1433–1447. https://doi.org/10.1109/TASE.2016.2603781

Feng, L., Qi, M., Hua, S., & Zhou, Q. (2019). Picking Station Location in Traditional and Flying-V Aisle Warehouses for Robotic Mobile Fulfillment System. In IEEE International Conference on Industrial Engineering and Engineering Management (Vols. 2019-Decem). https://doi.org/10.1109/IEEM.2018.8607301

Ferrara, A., Gebennini, E., & Grassi, A. (2014). Fleet sizing of laser guided vehicles and pallet shuttles in automated warehouses. International Journal of Production Economics, 157(1), 7–14. https://doi.org/10.1016/j.ijpe.2014.06.008

Füßler, D., Boysen, N., & Stephan, K. (2019). Trolley line picking: storage assignment and order sequencing to increase picking performance. OR Spectrum, 41(4), 1087–1121. https://doi.org/10.1007/s00291-019-00566-9

Hamann, H., Markarian, C., Auf Der Heide, F. M., & Wahby, M. (2018). Pick, Pack, & Survive: Charging Robots in a Modern Warehouse based on Online Connected Dominating Sets. Leibniz International Proceedings in Informatics, LIPIcs, 100, 221–2213. https://doi.org/10.4230/LIPIcs.FUN.2018.22

He, Z., Aggarwal, V., & Nof, S. Y. (2018). Differentiated service policy in smart warehouse automation. International Journal of Production Research, 56(22), 6956–6970. https://doi.org/10.1080/00207543.2017.1421789

Kattepur, A., Rath, H. K., Simha, A., & Mukherjee, A. (2018). Distributed optimization in multi-agent robotics for industry 4.0 warehouses. Proceedings of the ACM Symposium on Applied Computing, 808–815. https://doi.org/10.1145/3167132.3167221

Lamballais, T., Roy, D., & De Koster, R. B. M. (2017). Estimating performance in a Robotic Mobile Fulfillment System. European Journal of Operational Research, 256(3), 976–990. https://doi.org/10.1016/j.ejor.2016.06.063

Lamballais, T., Roy, D., & De Koster, R. B. M. (2020). Inventory allocation in robotic mobile fulfillment systems. IISE Transactions, 52(1), 1–17. https://doi.org/10.1080/24725854.2018.1560517

Lau, H. Y. K., Wong, V. W. K., & Lee, I. S. K. (2007). Immunity-based autonomous guided vehicles control. Applied Soft Computing Journal, 7(1), 41–57. https://doi.org/10.1016/j.asoc.2005.02.003

Le-Anh, T., & De Koster, R. B. M. (2006). A review of design and control of automated guided vehicle systems. European Journal of Operational Research, 171(1), 1–23. https://doi.org/10.1016/j.ejor.2005.01.036

Lee, C. K. M., Keung, K. L., Ng, K. K. H., & Lai, D. C. P. (2019). Simulation-based Multiple Automated Guided Vehicles Considering Charging and Collision-free Requirements in Automatic Warehouse. IEEE International Conference on Industrial Engineering and Engineering Management, 2019-Decem, 1376–1380. https://doi.org/10.1109/IEEM.2018.8607396

Lee, H. Y., & Murray, C. C. (2019). Robotics in order picking: evaluating warehouse layouts for pick, place, and transport vehicle routing systems. International

Journal of Production Research, 57(18), 5821–5841. https://doi.org/10.108
0/00207543.2018.1552031

Li, Z. P., Zhang, J. L., Zhang, H. J., & Hua, G. W. (2017). Optimal selection
of movable shelves under cargo-to-person picking mode. International
Journal of Simulation Modelling, 16(1), 145–156. https://doi.org/10.2507/
IJSIMM16(1)CO2

Lienert, T., Stigler, L., & Fottner, J. (2019). Failure-Handling Strategies For
Mobile Robots In Automated Warehouses. ECMS 2019 Proceedings Edited
by Mauro Iacono, Francesco Palmieri, Marco Gribaudo, Massimo Ficco,
199–205. https://doi.org/10.7148/2019-0199

Liu, Yiming, Chen, M., & Huang, H. (2019a). Multi-agent Pathfinding Based on
Improved Cooperative A* in Kiva System. In 2019 5th International
Conference on Control, Automation and Robotics, ICCAR 2019. https://
doi.org/10.1109/ICCAR.2019.8813319

Liu, Yubang, Ji, S., Su, Z., & Guo, D. (2019b). Multi-objective AGV scheduling
in an automatic sorting system of an unmanned (intelligent) warehouse by
using two adaptive genetic algorithms and a multi-adaptive genetic algorithm.
PLoS ONE, 14(12). https://doi.org/10.1371/journal.pone.0226161

Ly, G. B. (2019). Storage Assignment Policy and Route Planning of AGVS in
Warehouse Optimization. In Proceedings of 2019 International Conference on
System Science and Engineering, ICSSE 2019.

Merschformann, M., Lamballais, T., De Koster, R. B. M., & Suhl, L. (2019).
Decision rules for robotic mobile fulfillment systems. Operations Research
Perspectives, 6. https://doi.org/10.1016/j.orp.2019.100128

Polten, L., & Emde, S. (2020). Scheduling automated guided vehicles in very narrow aisle warehouses. Omega (United Kingdom). https://doi.org/10.1016/j.
omega.2020.102204

Roy, D., Nigam, S., De Koster, R. B. M., Adan, I., & Resing, J. (2019). Robot-
storage zone assignment strategies in mobile fulfillment systems. Transportation
Research Part E: Logistics and Transportation Review, 122, 119–142. https://
doi.org/10.1016/j.tre.2018.11.005

Rushton, A., Croucher, P., & Baker, P. (2014). Handbook of logistics and distribution management.

Sarkar, C., & Agarwal, M. (2019). Cannot avoid penalty? Let's minimize. In
Proceedings - IEEE International Conference on Robotics and Automation
(Vols. 2019-May). https://doi.org/10.1109/ICRA.2019.8794338

Sarkar, C., Paul, H. S., & Pal, A. (2018). A Scalable Multi-Robot Task Allocation
Algorithm. In Proceedings - IEEE International Conference on Robotics and
Automation. https://doi.org/10.1109/ICRA.2018.8460886

Sartoretti, G., Kerr, J., Shi, Y., Wagner, G., Satish Kumar, T. K., Koenig, S., &
Choset, H. (2019). PRIMAL: Pathfinding via Reinforcement and Imitation
Multi-Agent Learning. In IEEE Robotics and Automation Letters (Vol. 4,
Issue 3). https://doi.org/10.1109/LRA.2019.2903261

Semwal, T., Jha, S. S., & Nair, S. B. (2018). On Ordering Multi-Robot Task Executions within a Cyber Physical System. ACM Transactions on Autonomous and Adaptive Systems, 12(4), 1–27. https://doi.org/10.1145/3124677

Tai, R., Wang, J., Tian, W., Chen, W., Wang, H., & Zhou, Y. (2018). A Time-Efficient Approach to Solve Conflicts and Deadlocks for Scheduling AGVs in Warehousing Applications. 2018 IEEE International Conference on Real-Time Computing and Robotics (RCAR), 166–171. https://doi.org/10.1109/RCAR.2018.8621773

Vis, I. F. A. (2006). Survey of research in the design and control of automated guided vehicle systems. European Journal of Operational Research, 170(3), 677–709. https://doi.org/10.1016/j.ejor.2004.09.020

Vivaldini, K., Rocha, L. F., Martarelli, N. J., Becker, M., & Moreira, A. P. (2016). Integrated tasks assignment and routing for the estimation of the optimal number of AGVS. International Journal of Advanced Manufacturing Technology, 82(1–4), 719–736. https://doi.org/10.1007/s00170-015-7343-4

Wang, W., Wu, Y., Zheng, J., & Chi, C. (2020). A comprehensive framework for the design of modular robotic mobile fulfillment systems. IEEE Access, 8, 13259–13269. https://doi.org/10.1109/ACCESS.2020.2966403

Weidinger, F., Boysen, N., & Briskorn, D. (2018). Storage assignment with rack-moving mobile robots in KIVA warehouses. Transportation Science, 52(6), 1479–1495. https://doi.org/10.1287/trsc.2018.0826

Wior, I., Jerenz, S., & Fay, A. (2018). Automated transportation systems subject to interruptions in production and intralogistics - a survey and evaluation. International Journal of Logistics Systems and Management, 30(4), 421. https://doi.org/10.1504/IJLSM.2018.093581

Witczak, M., Majdzik, P., Stetter, R., & Lipiec, B. (2020). A fault-tolerant control strategy for multiple automated guided vehicles. Journal of Manufacturing Systems, 55, 56–68. https://doi.org/10.1016/j.jmsy.2020.02.009

Xu, J., Wang, J., & Chen, W. (2019). An efficient recharging task planning method for multi-robot autonomous recharging problem. In IEEE International Conference on Robotics and Biomimetics, ROBIO 2019. https://doi.org/10.1109/ROBIO49542.2019.8961439

Xue, F., & Dong, T. (2018). Research on the logistics robot task allocation method based on improved ant colony algorithm. In Int. J. Wireless and Mobile Computing (Vol. 14, Issue 4).

Yan, R. D., Dunnett, S. J., & Jackson, L. M. (2018). Optimising the maintenance strategy for a multi-AGV system using genetic algorithms. In Safety and Reliability – Safe Societies in a Changing World. CRC Press. https://doi.org/10.1201/9781351174664

Yan, R., Jackson, L. M., & Dunnett, S. J. (2017). Automated guided vehicle mission reliability modelling using a combined fault tree and Petri net approach. International Journal of Advanced Manufacturing Technology, 92(5–8), 1825–1837. https://doi.org/10.1007/s00170-017-0175-7

Yoshitake, H., Kamoshida, R., & Nagashima, Y. (2019). New Automated Guided Vehicle System Using Real-Time Holonic Scheduling for Warehouse Picking. IEEE Robotics and Automation Letters, 4(2), 1045–1052. https://doi. org/10.1109/LRA.2019.2894001

Yuan, R., Wang, H., & Li, J. (2019). The Pod Assignment Model and Algorithm in Robotic Mobile Fulfillment Systems. In Proceedings - IEEE International Conference on Service Operations and Logistics, and Informatics 2019, SOLI 2019. https://doi.org/10.1109/SOLI48380.2019.8955103

Yuan, Z., & Gong, Y. Y. (2017). Bot-in-time delivery for robotic mobile fulfillment systems. IEEE Transactions on Engineering Management, 64(1), 83–93. https://doi.org/10.1109/TEM.2016.2634540

Zhang, Z., Guo, Q., Chen, J., & Yuan, P. (2018). Collision-Free Route Planning for Multiple AGVs in an Automated Warehouse Based on Collision Classification. IEEE Access, 6, 26022–26035. https://doi.org/10.1109/ACCESS.2018.2819199

Zou, B., Xu, X., Gong, Y. (Yale), & De Koster, R. B. M. (2018). Evaluating battery charging and swapping strategies in a robotic mobile fulfillment system. European Journal of Operational Research, 267(2), 733–753. https://doi. org/10.1016/j.ejor.2017.12.008

Zou, Y., Zhang, D., & Qi, M. (2019). Order picking system optimization based on picker-robot collaboration. ACM International Conference Proceeding Series, 1–6. https://doi.org/10.1145/3364335.3364386

CHAPTER 6

Operational Decisions in Mobile Robot Automation

Operational decisions could be altered daily, and their effect on the warehouse operations could be observed within the same week or even within the same day. Thus, warehouse managers can be experimental with their decisions to optimise the operational performance objectives.

6.1 Mobile Robot Task Allocation

The purpose of allocating tasks to a mobile robot is to transport loads as swiftly as possible. Tasks could be generalised to operations such as replenishment, picking, and put-away tasks. Yet, the details of these tasks would differ according to the chosen mobile robot system. For instance, for barcode-guided mobile robot systems, a picking task could be the transportation of a shelf, while it could mean picking a box from a shelf for mobile picking robots. In either application, robots should divide these tasks into subtasks such as navigating to the correct location, performing the lifting or the picking operation, and then carrying the material to where it is supposed to go (Enright and Wurman 2011). Furthermore, there is a robot-specific task type about the energy management strategy. This wide range of tasks should be considered jointly, and the allocation of these tasks to robots requires efficiency optimisation models accordingly. As a general solution approach, few papers apply the queueing theory to

A. Yildirim et al., *Mobile Robot Automation in Warehouses*, Palgrave Studies in Logistics and Supply Chain Management, https://doi.org/10.1007/978-3-031-12307-8_6

sort tasks and observe their performance outputs (Lamballais et al. 2017; Wang et al. 2020).

Task allocation can be static/fixed or dynamic/online (Claes et al. 2017; Le-Anh et al. 2010; Le-Anh and De Koster 2006; Singhal et al. 2018; Vivaldini et al. 2015).

6.1.1   Static/Fixed Task Allocation

Tasks are grouped with a cut-off point in advance, a globally optimum plan is generated through an algorithm, and tasks are distributed to mobile robots (Le-Anh and De Koster 2006; Singhal et al. 2018). As this plan seeks the optimal solution, all information is gathered into a single source, and, for that reason, these solutions generally use centralised robot coordination (Fan et al. 2018; Singhal et al. 2018). However, the solution could be disrupted through unforeseen events such as breakdowns of robots, cancellation of tasks, or alterations in navigation times due to conflicts or deadlocks (Le-Anh and De Koster 2006; Vivaldini et al. 2015). Furthermore, a globally optimised solution might not be scalable to hundreds of robots as it requires centralised coordination, and having too many tasks in the system would require high computation time (Claes et al. 2017).

6.1.2   Dynamic/Online Task Allocation

To overcome the disadvantages of static task allocation, decision-makers began implementing dynamic task allocation solutions. Task allocation request in this system is triggered through system status changes such as completion of a task or arrival of a new task, and it is re-planned each time there is a trigger (Claes et al. 2017; Singhal et al. 2018; Vivaldini et al. 2015). With the help of this strategy, the system becomes more robust and scalable while considering prioritisation of tasks and alterations required due to unforeseen events (Le-Anh and De Koster 2006; Vivaldini et al. 2015). As these systems generally distribute the responsibility to intelligent robots with imperfect information and could only compute suboptimal solutions, studies focus on having the best performance on optimality/computation efficiency ratio (Ghassemi and Chowdhury 2018).

There are rules, exact approaches, heuristics, meta-heuristics, and artificial intelligence-based solutions for efficient task allocation of mobile robots (Confessore et al. 2013; De Koster et al. 2004; Fazlollahtabar and

Saidi-Mehrabad 2013; Le-Anh and De Koster 2006; Vis 2006). However, for simplicity, regardless of the warehousing operation and any general categorisation, Table 6.1 summarises task allocation approaches of empirical papers on task allocation. Later on, Chap. 7 mentions some algorithms in detail.

To conclude, dynamic task allocation approaches need to be adapted and verified through many robots as there might be hundreds of robots in the same warehouse, which increases the computational power required to allocate tasks and occurrences of unforeseen events.

6.2   Path Planning of Mobile Robots

Path planning concerns deciding the route mobile robots follow from the initial position to the desired position either to start or finish a task in the shortest time (Fazlollahtabar and Saidi-Mehrabad 2013). Generating the shortest path with numerous static and dynamic obstacles, including other mobile robots, makes path planning a significant concern. Algorithms are applied to generate paths from an existing location to the desired location in the shortest time. To assess the performance of these algorithms, Ng et al. (2020) mention four performance criteria: (1) completeness (the ability of the algorithm to find the complete path from source to destination), (2) optimality (finding the optimal path with the lowest time/cost), (3) time complexity (the computational time it takes to find the optimal path), and (4) space complexity (total computational memory consumed to obtain the optimal path).

There are multiple categorisations gathered from the literature for path planning algorithms. The first categorisation is based on the knowledge of the desired location relative to the existing location, which is divided as informed (with the map knowledge) and uninformed (without the map knowledge) search (Ng et al. 2020).

The second approach is to categorise path planning algorithms as static and dynamic (Abbas et al. 2018; Fazlollahtabar and Saidi-Mehrabad 2013; Vivaldini et al. 2015). In static algorithms, the shortest paths from the existing location to the desired location are previously defined and used, meaning minimum computation time and power (Abbas et al. 2018). However, it does not consider real data such as robot failures or traffic in a specific area which might affect the system performance as these algorithms depend on the travel time that is previously determined (Vivaldini et al. 2015). Dynamic algorithms, which act considering the real data,

Table 6.1   Task allocation approaches in the literature

Approach	Study	Approach	Study
Genetic algorithm	Dou et al. (2015), Kumar and Kumar (2018), Wang et al. (2019), Tsang et al. (2018) (+learning heuristic), Liu et al. (2019b) (multi adaptive), Zhang et al. (2019a) (+priority rule-based heuristic), Lee et al. (2020) (+auction-based winner determination)	Shortest distance (nearest vehicle first rule) / makespan	Liu et al. (2019c), Ly (2019), Qi et al. (2018), Tai et al. (2018, 2019), Xing et al. (2020), Jin et al. (2016) (+shortest task queue length of workstations), Rivas et al. (2019) (+auction)
Neighbourhood algorithms	Weidinger et al. (2018) (adaptive large neighbourhood search), Polten and Emde (2020) (lateness minimisation), Sarkar et al. (2018) (nearest neighbourhood-based clustering), Vivaldini et al. (2016) (+shortest job)	Artificial intelligence	Li et al. (2019a) (deep Q-network), Pagani et al. (2017) (neural networks trained with genetic algorithm), Boysen et al. (2017) (simulated annealing), Gunady et al. (2014) (reinforcement learning enhanced by state aggregation)
Shortest response time	Haiming et al. (2019), Li et al. (2019b) (idles vs closest)	Deadline-based	He et al. (2018) (+queueing solved by simulated annealing and order swapping), Sarkar and Agarwal (2019) (minimum penalty scheduling)
One out of N	Draganjac et al. (2016, 2020)	Auction	D'Emidio and Khan (2019), Liu et al. (2018), Singhal et al. (2018)
Linear programming	Lee and Murray (2019), Sabattini et al. (2017) (integer), Feng et al. (2019) (0–1 integer), Ono and Ishigami (2019) (mixed integer)	Negotiation	Kattepur et al. (2018) (contract net protocol), Digani et al. (2019) (waiting time minimisation)

(*continued*)

Table 6.1 (continued)

Approach	Study	Approach	Study
Greedy heuristic	Farinelli et al. (2017) (+distributed constraint optimisation), Claes et al. (2017) (+Monte Carlo tree search)	Prioritisation	Zhang et al. (2018), Füßler et al. (2019) (+order swapping)
Two-stage heuristic algorithm	Zou et al. (2019)	Heuristic distributed task allocation	Xu et al. (2019)
Mixed heuristic algorithm	Li et al. (2017)	Conflict-based min-cost-flow algorithm	Ma and Koenig (2016)
Time window-based dynamic routing	Smolic-Rocak et al. (2010)	Time-space network model optimisation	Yin and Xin (2019)
Sequentially execute interdependent tasks	Semwal et al. (2018)	Near-optimal assignment rule	Zou et al. (2017)
Artificial immune system-based exploration	Lau et al. (2007)	Token passing with task swaps	Ma et al. (2017)

Source: Authors

were developed to overcome this problem. Continuously searching for an optimum solution and having a high communication overhead, these systems are likely to demand computational power (Abbas et al. 2018).

The third categorisation is based on centralised and decentralised path planning algorithms (Draganjac et al. 2016; Fan et al. 2018). Centralised path planning relies on a central controller unit to dictate mobile robots by communicating with each one of them. Decentralised path planning does not require a central unit and all the information; instead, each robot is asked to compute its path to reduce the computational time (Fan et al. 2018).

The fourth categorisation is based on coupled and decoupled path planning algorithms (D'Emidio and Khan 2019; Dewangan et al. 2017; Draganjac et al. 2016; Sartoretti et al. 2019). Draganjac et al. (2016) and Fan et al. (2018) use coupled and decoupled path planning synonyms for centralised and decentralised approaches. Coupled path planning treats the multi-robot system as a single combined robot and cumulatively solves path planning (D'Emidio and Khan 2019). Coupled algorithms are high-dimensional algorithms that guarantee completeness and optimality, but they suffer from high computational performance requirements (D'Emidio and Khan 2019; Sartoretti et al. 2019). Decoupled path planning plans the paths of robots one by one and then adjusts these paths according to a prioritisation rule or through velocity adjustments to avoid robot colli-sions (Draganjac et al. 2016; Dewangan et al. 2017). This method decreases the computational performance requirement while risking the completeness and optimality criteria (D'Emidio and Khan 2019; Draganjac et al. 2016). Sartoretti et al. (2019) mention dynamically decoupled as a third category that exploits the advantages of coupled and decoupled algorithms by providing fast and complete solutions. We accept this cate-gorisation and include dynamically coupled algorithms.

Coupled algorithms such as Dijkstra's and A* consider the system as a whole and theoretically guarantee optimality and completeness, but they suffer from time and space complexities (Draganjac et al. 2016; Tai et al. 2018). Stern (2019) states these algorithms cannot be used, for instance, in a 500*500 grid environment with a large number of mobile robots. Further, coupled algorithms do not possess the ability to replan the paths in time as the environment changes (Ng et al. 2020).

Decoupled algorithms break path planning into instantaneous route planning and robot motion coordination forward through time to avoid collisions/conflicts. These algorithms use prioritisation to first deal with the most critical tasks (Draganjac et al. 2016; Tai et al. 2019). They are faster than coupled algorithms, but they suffer from completeness and optimality (Draganjac et al. 2016).

Dynamically coupled algorithms, a mixture of coupled and decoupled algorithms, plan routes and coordinate movement in the robot's local area to decrease time complexity while still possessing the ability to find opti-mal or near-optimal solutions (Sartoretti et al. 2019). Dynamically cou-pled algorithms could be further studied and combined with task allocation and conflict avoidance strategies to obtain resilient and sustainable systems.

Many algorithms are mentioned in review papers (Dewangan et al. 2017; Fazlollahtabar and Saidi-Mehrabad 2013; Stern 2019; Vis 2006; Vivaldini et al. 2015). Yet, we only summarise the algorithms that are applied in the empirical papers (Table 6.2).

Table 6.2 Path planning algorithms and studies

Approach	Study	Approach	Study
Improved A*	Krnjak et al. (2015) (+state lattice), Draganjac et al. (2016) (+state lattice), Santos et al. (2016) (Time enhanced A*), Zhang et al. (2019b), Digani et al. (2019) (+quadratic optimisation), Merschformann et al. (2019) (windowed hierarchical cooperative A*), Liu et al. (2019a) (improved cooperative A*), Liu et al. (2019c) (local cooperative A*), Lee et al. (2019b) (+Manhattan distance, route reservation, and turning times of robots)	**Shortest path algorithms such as Dijkstra's**	Yan et al. (2017), Smolic-Rocak et al. (2010), Lamballais et al. (2017, 2020), Farinelli et al. (2017), Sabattini et al. (2017), Ghassemi and Chowdhury (2018), Ono and Ishigami (2019), Draganjac et al. (2020), Li et al. (2020), Papcun et al. (2019), Bormann et al. (2019), Le-Anh et al. (2010), Vivaldini et al. (2016) (Enhanced Dijkstra's * turning times of robots), Zhang et al. (2018) (Improved Dijkstra's), Zou et al. (2018) (Shortest rectangular path), Hirayama and Nagao (2019) (+ dynamic path cost updates)
A*	Bechtsis et al. (2018), Digani et al. (2015, 2016), Haiming et al. (2019), Kumar and Kumar (2018), Wurman et al. (2008)	**Nearest neighbourhood**	Sarkar et al. (2018) (clustering + routing), Weidinger et al. (2018) (Adaptive large neighbourhood search), Zou et al. (2019) (+ two-stage heuristic algorithm)
k-shortest path planning	Li et al. (2019b), Ly (2019), Tai et al. (2018, 2019)	**Reinforcement learning**	Kamoshida and Kazama (2017) (Deep reinforcement learning), Sartoretti et al. (2019) (+imitation learning/ OD-recursive M*)

(*continued*)

Table 6.2 (continued)

Approach	Study	Approach	Study
Artificial potential field	Chen et al. (2018), Tsang et al. (2018) (+ recursive excitation/ relaxation)	Tabu search	Wei and Ni (2018) (tabu temporal difference +adaptive action selection and elimination), Xing et al. (2020) (+ mission group exchange with neighbourhood search)
Voronoi diagram	Petković et al. (2019), Piccinelli and Muradore (2018)	AD* (adaptive dynamic path-finding)	Ng et al. (2020)
Real-time holonic scheduling	Yoshitake et al. (2019)	Tracking fuzzy logic controller	Faisal et al. (2013)
PSO variants (Con-Per-PSO and SA-PSO)	Ma et al. (2014)	Q-learning	Dou et al. (2015)
Greedy algorithm and simulated annealing	Yuan et al. (2019)	Time-space network model optimisation	Yin and Xin (2019)
Monte Carlo tree search with iterative greedy heuristic	Claes et al. (2017)	Conflict-based min-cost-flow algorithm	Ma and Koenig (2016)
Safe interval path planning	D'Emidio and Khan (2019)	Insert route algorithm + solution improvement via late acceptance hill-climbing	Thanos et al. (2019)
Link weight increment heuristic with the time window	Fan et al. (2018)	Token passing with task swaps	Ma et al. (2017)
Linear programming	Lee and Murray (2019)	Incidental delivery	Liu et al. (2018)

(continued)

Table 6.2 (continued)

Approach	Study	Approach	Study
Genetic algorithm + auctions	Lee et al. (2020)	Oppositional-based learning + invasive weed optimisation	Panda et al. (2018)
D* with predictive control	Digani et al. (2015)		

Source: Authors

6.3   DEADLOCK RESOLUTION AND CONFLICT AVOIDANCE PLANS

Simply calculating a feasible route is not sufficient for multi-robot systems to operate. The operational structure should avoid or resolve any conflicts and deadlocks that may occur between mobile robots. This subject is often considered together with path planning under the name 'Multi-agent pathfinding' (Sartoretti et al. 2019; Stern 2019). Stern (2019) defines 'multi-agent pathfinding' as a problem of generating paths for multiple robots such that every robot reaches its destination and none of them has conflicts. Many papers address path planning and conflict avoidance subjects together in a combined approach (Chen et al. 2018; Ma et al. 2017; Ma and Koenig 2016; Merschformann et al. 2019; Panda et al. 2018; Santos et al. 2016; Thanos et al. 2019; Tsang et al. 2018; Wei and Ni 2018). Otherwise, a strategy to avoid conflicts and deadlocks should separately be planned (Vivaldini et al. 2015).

6.3.1   Conflicts

We divide types of conflicts into four, namely, cross, head-on, chasing, and stay-on conflicts (Lee et al. 2019b; Zhang et al. 2018) (Fig. 6.1). A cross conflict happens when two robots cross the intersection point at the same time, whereas a head-on conflict occurs when two robots facing each other try to go in the opposite directions. A chasing conflict happens when one mobile robot is following the same route with a slower robot that is in front of it. Finally, a stay-on conflict occurs when a robot tries to travel to a location that is occupied by a static (i.e. un/loading shelves or products) robot.

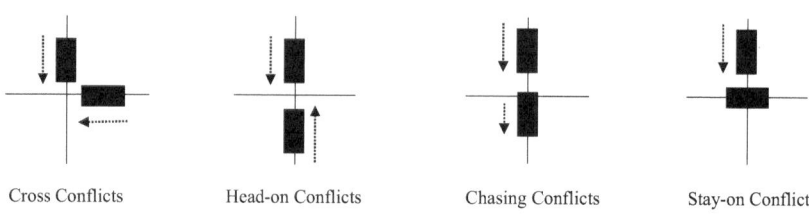

| Cross Conflicts | Head-on Conflicts | Chasing Conflicts | Stay-on Conflicts |

Fig. 6.1 Types of conflicts. (Adopted from: Lee et al. (2019b), Zhang et al. (2018))

Robots have a variety of actions to overcome conflicts. After a negotiation or due to a prioritisation strategy; a robot may wait, reroute, or step aside (Table 6.3). Routes of robots may also be converted to time windows, and the same actions can be taken according to detected overlaps (Tai et al. 2018). Finally, as Le-Anh and De Koster (2006) agree, having static or dynamic zones is an efficient method to avoid conflicts. In static zoning, the flow path of mobile robots is divided into several zones, and the traffic among the zones is tried to be balanced. In dynamic zoning, zone locations are not fixed and change according to the traffic flow (Digani et al. 2015; Le-Anh and De Koster 2006).

6.3.2   Deadlocks

Deadlocks can cause more time loss than conflicts if they cannot be immediately identified. Vis (2006) defines deadlocks as the inability of multiple robots to move further because each of them aims to occupy the location currently occupied by another robot in the same group. Many papers follow cycles to identify deadlocks and pause mobile robots, change a robot's task time to infinity, turn robots off for several seconds, or show neighbour robots as obstacles to force the robots to reroute (Qi et al. 2018; Sabattini et al. 2017; Santos et al. 2016; Smolic-Rocak et al. 2010; Yan et al. 2017).

A final remark is that most of the research is focussed on reactive approaches on conflict and deadlock avoidance which causes delays upon conflict. Instead, companies like Alibaba are adopting proactive approaches of finding an alternative route before a conflict occurs to maintain the flow of the robots (Lee et al. 2019a).

Table 6.3 Deadlock resolution and conflict avoidance strategies

Approach	Study	Approach	Study
Time windows	Smolic-Rocak et al. (2010) (+ changing task time to infinity for deadlocks), Vivaldini et al. (2016) (+route reservation), Santos et al. (2016) (A* with time windows + turning neighbour robots into obstacles), Fan et al. (2018) (+waiting), Lienert et al. (2019) (+route reservation), Tai et al. (2018, 2019) (+delays), Haiming et al. (2019) (+route reservation), Ly (2019) (+wait or reroute)	**Zones**	Confessore et al. (2013) (limits # of robots in zones and talks to robots about congested zones), Gunady et al. (2014), Digani et al. (2016, 2019); Lamballais et al. (2017), Qi et al. (2018), Li et al. (2019b) (+deadlock control algorithm), Zou et al. (2019), Digani et al. (2015) (zoning with voronoi decomposition), Liu et al. (2019c) (+maximum allowable # of robots)
Prioritisation and waiting or rerouting	Draganjac et al. (2016) (+private zone), Draganjac et al. 2020, Digani et al. (2016), Dou et al. (2015) (+learning reward), D'Emidio and Khan (2019), Qi et al. (2018) (first come first served), Polten and Emde (2020), Tai et al. (2019), Farinelli et al. (2017), Digani et al. (2015) (+negotiation)	**Check before each iteration: wait, swap, or reroute/ re-dispatch**	Krnjak et al. (2015), Qi et al. (2018), Kumar and Kumar (2018), Zhang et al. (2018), Xing et al. (2020), Singhal et al. (2018) (+congestion tracker), Tai et al. (2019)
Uni-directional aisles	Zou et al. (2017, 2018), Lamballais et al. (2020)	**Wait or reroute before starting**	Yin and Xin (2019), Zhang et al. (2018), Lee et al. (2019b)

(*continued*)

Table 6.3 (continued)

Approach	Study	Approach	Study
Velocity-based	Piccinelli and Muradore (2018)	Dynamic weight guidance	Liu et al. (2019a)
Conflict-based algorithms	Ma and Koenig (2016) (min cost-flow algorithm), Sabattini et al. (2017) (linear programming + conflict graph + making robots offline for 20 seconds for deadlocks), Liu et al. (2019c)	Rule-based	Yan et al. (2017), Hirayama and Nagao (2019) (+ treat intersection as static points)
A cost incurred if a collision occurs	Jin et al. (2016), Sartoretti et al. (2019)	Artificial potential function	Tsang et al. (2018) (+ recursive excitation/ relaxation), Chen et al. (2018) (improved artificial potential function)
Deep Q-network (reinforcement learning) rewards for staying away from the traffic	Li et al. (2019a)	Gaussian mixture-based background segmentation algorithm for detection + reroute	Ng et al. (2020)
Obstacles avoiding fuzzy logic controller	Faisal et al. (2013)	PSO variants (Con-Per-PSO and SA-PSO)	Ma et al. (2014)
Oppositional-based learning + invasive weed optimisation	Panda et al. (2018)	Insert route algorithm + improvement via late acceptance hill-climbing	Thanos et al. (2019)
Windowed hierarchical cooperative A*	Merschformann et al. (2019)	Tabu temporal learning with adaptive action selection and elimination	Wei and Ni (2018)
Token passing with task swaps	Ma et al. (2017)		

REFERENCES

Abbas, A. S., Mohamed, T. A., & Hazem, M. (2018). Optimization of Warehouse Material Handling Parameters to Enhance the Efficiency of Automated Sorting and Storage Systems. 11(1), 2018.

Bechtsis, D., Tsolakis, N., Vlachos, D., & Srai, J. S. (2018). Intelligent Autonomous Vehicles in digital supply chains: A framework for integrating innovations towards sustainable value networks. Journal of Cleaner Production, 181, 60–71. https://doi.org/10.1016/j.jclepro.2018.01.173

Bormann, R., de Brito, B. F., Lindermayr, J., Omainska, M., & Patel, M. (2019). Towards Automated Order Picking Robots for Warehouses and Retail. Lecture Notes in Computer Science (Including Subseries Lecture Notes in Artificial Intelligence and Lecture Notes in Bioinformatics), 11754 LNCS, 185–198. https://doi.org/10.1007/978-3-030-34995-0_18

Boysen, N., Briskorn, D., & Emde, S. (2017). Parts-to-picker based order processing in a rack-moving mobile robots environment. European Journal of Operational Research, 262(2), 550–562. https://doi.org/10.1016/j.ejor.2017.03.053

Chen, H., Wang, Q., Yu, M., Cao, J., & Sun, J. (2018). Path Planning for Multi-robot Systems in Intelligent Warehouse. Lecture Notes in Computer Science (Including Subseries Lecture Notes in Artificial Intelligence and Lecture Notes in Bioinformatics), 11226 LNCS, 148–159. https://doi.org/10.1007/978-3-030-02738-4_13

Claes, D., Oliehoek, F., Baier, H., & Tuyls, K. (2017). Decentralised Online Planning for Multi-Robot Warehouse Commissioning.

Confessore, G., Fabiano, M., & Liotta, G. (2013). A network flow based heuristic approach for optimising AGV movements. Journal of Intelligent Manufacturing, 24(2), 405–419. https://doi.org/10.1007/s10845-011-0612-7

D'Emidio, M., & Khan, I. (2019). Collision-free allocation of temporally constrained tasks in multi-robot systems. Robotics and Autonomous Systems, 119, 151–172. https://doi.org/10.1016/j.robot.2019.07.002

De Koster, R. B. M., Le-Anh, T., & Van Der Meer, J. R. (2004). Testing and classifying vehicle dispatching rules in three real-world settings. Journal of Operations Management, 22(4 SPEC. ISS.), 369–386. https://doi.org/10.1016/j.jom.2004.05.006

Dewangan, R. K., Shukla, A., & Godfrey, W. W. (2017). Survey on prioritized multi robot path planning. In 2017 IEEE International Conference on Smart Technologies and Management for Computing, Communication, Controls, Energy and Materials, ICSTM 2017 - Proceedings. https://doi.org/10.1109/ICSTM.2017.8089197

Digani, V., Hsieh, M. A., Sabattini, L., & Secchi, C. (2019). Coordination of multiple AGVs: a quadratic optimization method. Autonomous Robots, 43(3), 539–555. https://doi.org/10.1007/s10514-018-9730-9

Digani, V., Sabattini, L., & Secchi, C. (2016). A probabilistic Eulerian traffic model for the coordination of multiple AGVs in automatic warehouses. IEEE Robotics and Automation Letters, 1(1), 26–32. https://doi.org/10.1109/LRA.2015.2505646

Digani, V., Sabattini, L., Secchi, C., & Fantuzzi, C. (2015). Ensemble Coordination Approach in Multi-AGV Systems Applied to Industrial Warehouses. IEEE Transactions on Automation Science and Engineering, 12(3), 922–934. https://doi.org/10.1109/TASE.2015.2446614

Dou, J., Chen, C., & Yang, P. (2015). Genetic Scheduling and Reinforcement Learning in Multirobot Systems for Intelligent Warehouses. Mathematical Problems in Engineering, 2015. https://doi.org/10.1155/2015/597956

Draganjac, I., Miklic, D., Kovacic, Z., Vasiljevic, G., & Bogdan, S. (2016). Decentralized Control of Multi-AGV Systems in Autonomous Warehousing Applications. IEEE Transactions on Automation Science and Engineering, 13(4), 1433–1447. https://doi.org/10.1109/TASE.2016.2603781

Draganjac, I., Petrović, T., Miklić, D., Kovačić, Z., & Oršulić, J. (2020). Highly-scalable traffic management of autonomous industrial transportation systems. Robotics and Computer-Integrated Manufacturing, 63. https://doi.org/10.1016/j.rcim.2019.101915

Enright, J. J., & Wurman, P. R. (2011). Optimization and coordinated autonomy in mobile fulfillment systems. AAAI Workshop - Technical Report, WS-11-09, 33–38.

Faisal, M., Hedjar, R., Al Sulaiman, M., & Al-Mutib, K. (2013). Fuzzy logic navigation and obstacle avoidance by a mobile robot in an unknown dynamic environment. International Journal of Advanced Robotic Systems, 10. https://doi.org/10.5772/54427

Fan, Z., Gu, C., Yin, X., Liu, C., & Huang, H. (2018). Time window based path planning of multi-AGVs in logistics center. Proceedings - 2017 10th International Symposium on Computational Intelligence and Design, ISCID 2017, 2, 161–166. https://doi.org/10.1109/ISCID.2017.40

Farinelli, A., Boscolo, N., Zanotto, E., & Pagello, E. (2017). Advanced approaches for multi-robot coordination in logistic scenarios. Robotics and Autonomous Systems, 90, 34–44. https://doi.org/10.1016/j.robot.2016.08.010

Fazlollahtabar, H., & Saidi-Mehrabad, M. (2013). Methodologies to Optimize Automated Guided Vehicle Scheduling and Routing Problems: A Review Study. Journal of Intelligent and Robotic Systems: Theory and Applications, 77(3–4), 525–545. https://doi.org/10.1007/s10846-013-0003-8

Feng, L., Qi, M., Hua, S., & Zhou, Q. (2019). Picking Station Location in Traditional and Flying-V Aisle Warehouses for Robotic Mobile Fulfillment

System. In IEEE International Conference on Industrial Engineering and Engineering Management (Vols. 2019-Decem). https://doi.org/10.1109/IEEM.2018.8607301

Füßler, D., Boysen, N., & Stephan, K. (2019). Trolley line picking: storage assignment and order sequencing to increase picking performance. OR Spectrum, 41(4), 1087–1121. https://doi.org/10.1007/s00291-019-00566-9

Ghassemi, P., & Chowdhury, S. (2018). Decentralized task allocation in multi-robot systems via bipartite graph matching augmented with fuzzy clustering. In ASME.

Gunady, M. K., Gomaa, W., & Takeuchi, I. (2014). Aggregate Reinforcement Learning for multi-agent territory division: The Hide-and-Seek game. Engineering Applications of Artificial Intelligence, 34, 122–136. https://doi.org/10.1016/j.engappai.2014.05.012

Haiming, L., Weidong, L., Mei, Z., & An, C. (2019). Algorithm of path planning based on time window for multiple mobile robots in warehousing system. In Chinese Control Conference, CCC (Vols. 2019-July). https://doi.org/10.23919/ChiCC.2019.8866325

He, Z., Aggarwal, V., & Nof, S. Y. (2018). Differentiated service policy in smart warehouse automation. International Journal of Production Research, 56(22), 6956–6970. https://doi.org/10.1080/00207543.2017.1421789

Hirayama, C., & Nagao, T. (2019). Dynamic path costs update method reflecting delivery tendencies for multi-agent delivery tasks. Conference Proceedings - IEEE International Conference on Systems, Man and Cybernetics, 2019-Octob, 4361–4366. https://doi.org/10.1109/SMC.2019.8914509

Jin, X., Zhong, M., Quan, X., Li, S., & Zhang, H. (2016). Dynamic scheduling of mobile-robotic warehouse logistics system. In Chinese Control Conference, CCC (Vols. 2016-Augus). https://doi.org/10.1109/ChiCC.2016.7553799

Kamoshida, R., & Kazama, Y. (2017). Acquisition of Automated Guided Vehicle Route Planning Policy Using Deep Reinforcement Learning.

Kattepur, A., Rath, H. K., Simha, A., & Mukherjee, A. (2018). Distributed optimization in multi-agent robotics for industry 4.0 warehouses. Proceedings of the ACM Symposium on Applied Computing, 808–815. https://doi.org/10.1145/3167132.3167221

Krnjak, A., Draganjac, I., Bogdan, S., Petrovic, T., Miklic, D., & Kovacic, Z. (2015). Decentralized control of free ranging AGVs in warehouse environments. In Proceedings - IEEE International Conference on Robotics and Automation (Vols. 2015-June, Issue June). https://doi.org/10.1109/ICRA.2015.7139465

Kumar, N. V., & Kumar, C. S. (2018). Development of collision free path planning algorithm for warehouse mobile robot. Procedia Computer Science, 133, 456–463. https://doi.org/10.1016/j.procs.2018.07.056

Lamballais, T., Roy, D., & De Koster, R. B. M. (2017). Estimating performance in a Robotic Mobile Fulfillment System. European Journal of Operational Research, 256(3), 976–990. https://doi.org/10.1016/j.ejor.2016.06.063

Lamballais, T., Roy, D., & De Koster, R. B. M. (2020). Inventory allocation in robotic mobile fulfillment systems. IISE Transactions, 52(1), 1–17. https://doi.org/10.1080/24725854.2018.1560517

Lau, H. Y. K., Wong, V. W. K., & Lee, I. S. K. (2007). Immunity-based autonomous guided vehicles control. Applied Soft Computing Journal, 7(1), 41–57. https://doi.org/10.1016/j.asoc.2005.02.003

Le-Anh, T., & De Koster, R. B. M. (2006). A review of design and control of automated guided vehicle systems. European Journal of Operational Research, 171(1), 1–23. https://doi.org/10.1016/j.ejor.2005.01.036

Le-Anh, T., De Koster, R. B. M., & Yu, Y. (2010). Performance evaluation of dynamic scheduling approaches in vehicle-based internal transport systems. International Journal of Production Research, 48(24), 7219–7242. https://doi.org/10.1080/00207540903443279

Lee, C. K. M., Keung, K. L., Ng, K. K. H., & Lai, D. C. P. (2019a). Simulation-based Multiple Automated Guided Vehicles Considering Charging and Collision-free Requirements in Automatic Warehouse. IEEE International Conference on Industrial Engineering and Engineering Management, 2019-Decem, 1376–1380. https://doi.org/10.1109/IEEM.2018.8607396

Lee, C., Lin, B., Ng, K. K. H., Lv, Y., & Tai, W. C. (2019b). Smart robotic mobile fulfillment system with dynamic conflict-free strategies considering cyber-physical integration. Advanced Engineering Informatics, 42. https://doi.org/10.1016/j.aei.2019.100998

Lee, C. W., Wong, W. P., Ignatius, J., Rahman, A., & Tseng, M. L. (2020). Winner determination problem in multiple automated guided vehicle considering cost and flexibility. Computers and Industrial Engineering, 142. https://doi.org/10.1016/j.cie.2020.106337

Lee, H. Y., & Murray, C. C. (2019). Robotics in order picking: evaluating warehouse layouts for pick, place, and transport vehicle routing systems. International Journal of Production Research, 57(18), 5821–5841. https://doi.org/10.1080/00207543.2018.1552031

Li, M. P., Sankaran, P., Kuhl, M. E., Ptucha, R., Ganguly, A., & Kwasinski, A. (2019a). Task Selection by Autonomous Mobile Robots in A Warehouse Using Deep Reinforcement Learning. In Proceedings - Winter Simulation Conference (Vols. 2019-Decem). https://doi.org/10.1109/WSC40007.2019.9004792

Li, X., Zhang, C., Yang, W., & Qi, M. (2019b). Multi-AGVs conflict-free routing and dynamic dispatching strategies for automated warehouses. Lecture Notes in Electrical Engineering, 513, 277–286. https://doi.org/10.1007/978-981-13-1059-1_26

Li, Z., Barenji, A. V., Jiang, J., Zhong, R. Y., & Xu, G. (2020). A mechanism for scheduling multi robot intelligent warehouse system face with dynamic demand. Journal of Intelligent Manufacturing, 31(2), 469–480. https://doi.org/10.1007/s10845-018-1459-y

Li, Z. P., Zhang, J. L., Zhang, H. J., & Hua, G. W. (2017). Optimal selection of movable shelves under cargo-to-person picking mode. International Journal of Simulation Modelling, 16(1), 145–156. https://doi.org/10.2507/IJSIMM16(1)CO2

Lienert, T., Stigler, L., & Fottner, J. (2019). Failure-Handling Strategies For Mobile Robots In Automated Warehouses. ECMS 2019 Proceedings Edited by Mauro Iacono, Francesco Palmieri, Marco Gribaudo, Massimo Ficco, 199–205. https://doi.org/10.7148/2019-0199

Liu, Yiming, Chen, M., & Huang, H. (2019a). Multi-agent Pathfinding Based on Improved Cooperative A* in Kiva System. In 2019 5th International Conference on Control, Automation and Robotics, ICCAR 2019. https://doi.org/10.1109/ICCAR.2019.8813319

Liu, Yubang, Ji, S., Su, Z., & Guo, D. (2019b). Multi-objective AGV scheduling in an automatic sorting system of an unmanned (intelligent) warehouse by using two adaptive genetic algorithms and a multi-adaptive genetic algorithm. PLoS ONE, 14(12). https://doi.org/10.1371/journal.pone.0226161

Liu, Z., Wang, H., Chen, W., & Liu, Y. H. (2018). Distributed pair-wised transportation planning with incidental deliveries for multiple mobile robots. In 2017 IEEE International Conference on Real-Time Computing and Robotics, RCAR 2017 (Vols. 2017-July). https://doi.org/10.1109/RCAR.2017.8311859

Liu, Z., Zhou, S., Wang, H., Shen, Y., Li, H., & Liu, Y. H. (2019c). A hierarchical framework for coordinating large-scale robot networks. In Proceedings - IEEE International Conference on Robotics and Automation (Vols. 2019-May). https://doi.org/10.1109/ICRA.2019.8793719

Ly, G. B. (2019). Storage Assignment Policy and Route Planning of AGVS in Warehouse Optimization. In Proceedings of 2019 International Conference on System Science and Engineering, ICSSE 2019.

Ma, H., & Koenig, S. (2016). Optimal Target Assignment and Path Finding for Teams of Agents. Proceedings of the 15th International Conference on Autonomous Agents and Multiagent Systems.

Ma, H., Kumar, T. K. S., Li, J., & Koenig, S. (2017). Lifelong multi-Agent path finding for online pickup and delivery tasks. Proceedings of the International Joint Conference on Autonomous Agents and Multiagent Systems, AAMAS, 2, 837–845.

Ma, Y., Wang, H., Xie, Y., & Guo, M. (2014). Path planning for multiple mobile robots under double-warehouse. Information Sciences, 278, 357–379. https://doi.org/10.1016/j.ins.2014.03.058

Merschformann, M., Lamballais, T., De Koster, R. B. M., & Suhl, L. (2019). Decision rules for robotic mobile fulfillment systems. Operations Research Perspectives, 6. https://doi.org/10.1016/j.orp.2019.100128

Ng, M. K., Chong, Y. W., Ko, K. man, Park, Y. H., & Leau, Y. B. (2020). Adaptive path finding algorithm in dynamic environment for warehouse robot. Neural Computing and Applications. https://doi.org/10.1007/s00521-020-04764-3

Ono, Y., & Ishigami, G. (2019). Routing Problem of Multiple Mobile Robots with Human Workers for Pickup and Dispatch Tasks in Warehouse. Proceedings of the 2019 IEEE/SICE International Symposium on System Integration, SII 2019, 176–181. https://doi.org/10.1109/SII.2019.8700428

Pagani, P., Colling, D., & Furmans, K. (2017). Neural Network-Based Genetic Job Assignment for Automated Guided Vehicles. Logistics Journal, 2017. https://doi.org/10.2195/lj_Proc_pagani_en_201710_01

Panda, M. R., Das, P. K., Dutta, S., & Pradhan, S. K. (2018). Optimal path planning for mobile robots using oppositional invasive weed optimization. Computational Intelligence, 34(4), 1072–1100. https://doi.org/10.1111/coin.12166

Papcun, P., Cabadaj, J., Kajati, E., Romero, D., Landryova, L., Vascak, J., & Zolotova, I. (2019). Augmented Reality for Humans-Robots Interaction in Dynamic Slotting "Chaotic Storage" Smart Warehouses. IFIP Advances in Information and Communication Technology, 566, 633–641. https://doi.org/10.1007/978-3-030-30000-5_77

Petković, T., Puljiz, D., Marković, I., & Hein, B. (2019). Human intention estimation based on hidden Markov model motion validation for safe flexible robotized warehouses. Robotics and Computer-Integrated Manufacturing, 57, 182–196. https://doi.org/10.1016/j.rcim.2018.11.004

Piccinelli, N., & Muradore, R. (2018). Hybrid Motion Planner Integrating Global Voronoi Diagrams and Local Velocity Obstacle Method. In 2018 European Control Conference, ECC 2018. https://doi.org/10.23919/ECC.2018.8550127

Polten, L., & Emde, S. (2020). Scheduling automated guided vehicles in very narrow aisle warehouses. Omega (United Kingdom). https://doi.org/10.1016/j.omega.2020.102204

Qi, M., Li, X., Yan, X., & Zhang, C. (2018). On the evaluation of AGVS-based warehouse operation performance. Simulation Modelling Practice and Theory, 87, 379–394. https://doi.org/10.1016/j.simpat.2018.07.015

Rivas, D., Jiménez-Jané, J., & Ribas-Xirgo, L. (2019). Auction Model for Transport Order Assignment in AGV Systems. Advances in Intelligent Systems and Computing, 855, 227–241. https://doi.org/10.1007/978-3-319-99885-5_16

Sabattini, L., Digani, V., Secchi, C., & Fantuzzi, C. (2017). Optimized simultaneous conflict-free task assignment and path planning for multi-AGV systems. In IEEE International Conference on Intelligent Robots and Systems (Vols. 2017-Septe). https://doi.org/10.1109/IROS.2017.8202278

Santos, J., Costa, P., Rocha, L., Vivaldini, K., Moreira, A. P., & Veiga, G. (2016). Validation of a time based routing algorithm using a realistic automatic warehouse scenario. Advances in Intelligent Systems and Computing, 418, 81–92. https://doi.org/10.1007/978-3-319-27149-1_7

Sarkar, C., & Agarwal, M. (2019). Cannot avoid penalty? Let's minimize. In Proceedings - IEEE International Conference on Robotics and Automation (Vols. 2019-May). https://doi.org/10.1109/ICRA.2019.8794338

Sarkar, C., Paul, H. S., & Pal, A. (2018). A Scalable Multi-Robot Task Allocation Algorithm. In Proceedings - IEEE International Conference on Robotics and Automation. https://doi.org/10.1109/ICRA.2018.8460886

Sartoretti, G., Kerr, J., Shi, Y., Wagner, G., Satish Kumar, T. K., Koenig, S., & Choset, H. (2019). PRIMAL: Pathfinding via Reinforcement and Imitation Multi-Agent Learning. In IEEE Robotics and Automation Letters (Vol. 4, Issue 3). https://doi.org/10.1109/LRA.2019.2903261

Semwal, T., Jha, S. S., & Nair, S. B. (2018). On Ordering Multi-Robot Task Executions within a Cyber Physical System. ACM Transactions on Autonomous and Adaptive Systems, 12(4), 1–27. https://doi.org/10.1145/3124677

Singhal, A., Singh, H. V., Penumatsa, A., Bhatt, N., Ambwani, P., Kumar, S., & Sinha, R. (2018). An actor based architecture for multi-robot system with application to warehouse. IoPARTS 2018 - Proceedings of the 2018 International Workshop on Internet of People, Assistive Robots and ThingS, 13–18. https://doi.org/10.1145/3215525.3215530

Smolic-Rocak, N., Bogdan, S., Kovacic, Z., & Petrovic, T. (2010). Time windows based dynamic routing in multi-AGV systems. IEEE Transactions on Automation Science and Engineering, 7(1), 151–155. https://doi.org/10.1109/TASE.2009.2016350

Stern, R. (2019). Multi-agent path finding – an overview. Lecture Notes in Computer Science (Including Subseries Lecture Notes in Artificial Intelligence and Lecture Notes in Bioinformatics), 11866 LNAI, 96–115. https://doi.org/10.1007/978-3-030-33274-7_6

Tai, R., Wang, J., & Chen, W. (2019). A prioritized planning algorithm of trajectory coordination based on time windows for multiple AGVs with delay disturbance. Assembly Automation, 39(5), 753–768. https://doi.org/10.1108/AA-03-2019-0054

Tai, R., Wang, J., Tian, W., Chen, W., Wang, H., & Zhou, Y. (2018). A Time-Efficient Approach to Solve Conflicts and Deadlocks for Scheduling AGVs in Warehousing Applications. 2018 IEEE International Conference on Real-Time Computing and Robotics (RCAR), 166–171. https://doi.org/10.1109/RCAR.2018.8621773

Thanos, E., Wauters, T., & Vanden Berghe, G. (2019). Dispatch and conflict-free routing of capacitated vehicles with storage stack allocation. Journal of the Operational Research Society. https://doi.org/10.1080/01605682.2019.1595191

Tsang, K. F. E., Ni, Y., Wong, C. F. R., & Shi, L. (2018). A Novel Warehouse Multi-Robot Automation System with Semi-Complete and Computationally Efficient Path Planning and Adaptive Genetic Task Allocation Algorithms.

Vis, I. F. A. (2006). Survey of research in the design and control of automated guided vehicle systems. European Journal of Operational Research, 170(3), 677–709. https://doi.org/10.1016/j.ejor.2004.09.020

Vivaldini, K., Rocha, L. F., Becker, M., & Moreira, A. P. (2015). Comprehensive review of the dispatching, scheduling and routing of AGVs. Lecture Notes in Electrical Engineering, 321 LNEE, 505–514. https://doi.org/10.1007/978-3-319-10380-8_48

Vivaldini, K., Rocha, L. F., Martarelli, N. J., Becker, M., & Moreira, A. P. (2016). Integrated tasks assignment and routing for the estimation of the optimal number of AGVS. International Journal of Advanced Manufacturing Technology, 82(1–4), 719–736. https://doi.org/10.1007/s00170-015-7343-4

Wang, H., Chen, W., & Wang, J. (2019). Heterogeneous multi-agent routing strategy for robot-and-picker-to-good order fulfillment system. Advances in Intelligent Systems and Computing, 867, 237–249. https://doi.org/10.1007/978-3-030-01370-7_19

Wang, W., Wu, Y., Zheng, J., & Chi, C. (2020). A comprehensive framework for the design of modular robotic mobile fulfillment systems. IEEE Access, 8, 13259–13269. https://doi.org/10.1109/ACCESS.2020.2966403

Wei, C., & Ni, F. (2018). Tabu temporal difference learning for robot path planning in uncertain environments. Lecture Notes in Computer Science (Including Subseries Lecture Notes in Artificial Intelligence and Lecture Notes in Bioinformatics), 10965 LNAI, 123–134. https://doi.org/10.1007/978-3-319-96728-8_11

Weidinger, F., Boysen, N., & Briskorn, D. (2018). Storage assignment with rack-moving mobile robots in KIVA warehouses. Transportation Science, 52(6), 1479–1495. https://doi.org/10.1287/trsc.2018.0826

Wurman, P. R., DAndrea, R., & Mountz, M. (2008). Coordinating Hundreds of Cooperative, Autonomous Vehicles in Warehouses.

Xing, L., Liu, Y., Li, H., Wu, C. C., Lin, W. C., & Chen, X. (2020). A novel tabu search algorithm for multi-AGV routing problem. Mathematics, 8(2). https://doi.org/10.3390/math8020279

Xu, J., Wang, J., & Chen, W. (2019). An efficient recharging task planning method for multi robot autonomous recharging problem. In IEEE International Conference on Robotics and Biomimetics, ROBIO 2019. https://doi.org/10.1109/ROBIO49542.2019.8961439

Yan, X., Zhang, C., & Qi, M. (2017). Multi-AGVs Collision-Avoidance and Deadlock-Control for Item-to-Human Automated Warehouse. IEEE International Conference on Robotics and Biomimetics, ROBIO 2019.

Yin, S., & Xin, J. (2019). Path planning of multiple AGVs using a time-space network model. In Proceedings - 2019 34rd Youth Academic Annual Conference of Chinese Association of Automation, YAC 2019. https://doi.org/10.1109/YAC.2019.8787726

Yoshitake, H., Kamoshida, R., & Nagashima, Y. (2019). New Automated Guided Vehicle System Using Real-Time Holonic Scheduling for Warehouse Picking. IEEE Robotics and Automation Letters, 4(2), 1045–1052. https://doi.org/10.1109/LRA.2019.2894001

Yuan, R., Wang, H., & Li, J. (2019). The Pod Assignment Model and Algorithm in Robotic Mobile Fulfillment Systems. In Proceedings - IEEE International Conference on Service Operations and Logistics, and Informatics 2019, SOLI 2019. https://doi.org/10.1109/SOLI48380.2019.8955103

Zhang, J., Yang, F., & Weng, X. (2019a). A Building-Block-Based Genetic Algorithm for Solving the Robots Allocation Problem in a Robotic Mobile Fulfilment System. Mathematical Problems in Engineering, 2019, 1–15. https://doi.org/10.1155/2019/6153848

Zhang, Y., Li, L. L., Lin, H. C., Ma, Z., & Zhao, J. (2019b). Development of path planning approach using improved a-star algorithm in AGV system. Journal of Internet Technology, 20(3), 915–924. https://doi.org/10.3966/160792642019052003023

Zhang, Z., Guo, Q., Chen, J., & Yuan, P. (2018). Collision-Free Route Planning for Multiple AGVs in an Automated Warehouse Based on Collision Classification. IEEE Access, 6, 26022–26035. https://doi.org/10.1109/ACCESS.2018.2819199

Zou, B., Gong, Y. (Yale), Xu, X., & Yuan, Z. (2017). Assignment rules in robotic mobile fulfilment systems for online retailers. International Journal of Production Research, 55(20), 6175–6192. https://doi.org/10.1080/00207543.2017.1331050

Zou, B., Xu, X., Gong, Y. (Yale), & De Koster, R. B. M. (2018). Evaluating battery charging and swapping strategies in a robotic mobile fulfillment system. European Journal of Operational Research, 267(2), 733–753. https://doi.org/10.1016/j.ejor.2017.12.008

Zou, Y., Zhang, D., & Qi, M. (2019). Order picking system optimization based on picker-robot collaboration. ACM International Conference Proceeding Series, 1–6. https://doi.org/10.1145/3364335.3364386

CHAPTER 7

Managerial Decision Framework

This chapter merges the three primary decision levels (strategic, tactical, and operational) and their thirteen managerial focus areas to form the managerial decision framework (Fig. 7.1). Along with the focus areas, decision questions that should be answered while considering these focus areas and papers mentioning the focus areas are provided.

There is a precedence order for each focus area mentioned in the framework. For instance, it is impossible to determine the facility layout before making decisions on the mobile robot system. Similarly, before considering how to manage human-robot interaction, robot type should be determined as AGVs would need more attention and scenarios in terms of safety management compared to AMRs. As another example, decision-makers cannot decide how many robots they require in their operations before choosing their order management strategies.

The following subsections (7.1 and 7.2) address concepts that have not formally affected the managerial decision framework yet are eligible for further attention. They will provide a discussion on change management and a critique of algorithms that may be deployed to manage robots' behaviour in the warehouse.

A. Yildirim et al., *Mobile Robot Automation in Warehouses*, Palgrave Studies in Logistics and Supply Chain Management, https://doi.org/10.1007/978-3-031-12307-8_7

Decision Level		Managerial Focus	Decision Questions	References
Pre-implementation / *Long-term decisions*	**Strategic Level**	**Evaluation Criteria and Mobile Robot System Selection**	According to which criteria to choose a mobile robot system?	Azadeh et al. 2019; Huang et al. 2015; Boysen et al. 2019; Roy et al. 2019; Schmidt and Schulze 2009
		Identifying Key Performance Indicators	How to choose warehouse-specific and mobile robot system-specific KPIs?	Bechtsis et al., 2017; Dou et al., 2015; Yuan and Gong., 2017
		Type of Robots and Their Coordination	AGVs or AMRs? – Centralised or decentralised coordination?	Bechtsis et al., 2017; Draganjac et al., 2016; Krnjak et al., 2015; Zavadskas et al., 2018
		Facility Layout	How many floors? What is the flow-path of robots? From where to pick and drop products and refuel robots?	Le-Anh and De Koster, 2006; Merschformann et al., 2019; Vis, 2006
		Human-Robot Interaction Management	Which policies should be followed in task distribution? How to manage the change? How to create a safe environment?	Azadeh et al., 2019; Bechtsis et al., 2017; Boysen et al., 2019; Inam and Raizer, 2018; Moeller et al., 2016
Adjustable in post-implementation / *Short- and medium- term decisions*	**Tactical Level**	**Storage Assignment Plan**	What are the rules or approaches on storage? How many pick locations are required for an SKU?	Lamballais et al., 2020; Merschformann et al., 2019; Weidinger et al., 2018
		Order Management Plan	Online/dynamic or offline/static order management? Batches, waves, or clusters of orders?	Boysen et al., 2017, 2019; Merschformann et al., 2019; Zou et al., 2019
		Quantity of Robots	How to decide the number of robots in the operation?	Ferrara et al., 2014; Lamballais et al., 2017; Le-Anh and De Koster, 2006; Roy et al., 2019
		Maintenance and Failure Handling	How often to do maintenance and how to estimate robot failures? What to do if a robot fails?	Bechtsis et al., 2017; Lienert et al., 2019; Witczak et al., 2020; Yan, et al., 2017; Yan et al., 2018
		Robot Energy Management	When, where, and how to charge? Plug-in versus battery swap?	Hamann et al., 2018; Kattepur et al., 2018; Xu et al., 2019; Zou et al., 2018
	Operational Level	**Mobile Robot Task Allocation Plan**	Where to go? When to go? Pick or replenishment or charging order?	Claes et al., 2017; Le-Anh et al., 2010; Merschformann et al., 2019; Vivaldini et al., 2015
		Path Planning of Mobile Robots	How to go in the shortest time?	Fazlollahtabar and Saidi-Mehrabad, 2013; Ng et al., 2020; Sartoretti et al., 2019; Vis, 2006
		Deadlock Resolution and Conflict Avoidance Plans	How to overcome conflicts and deadlocks?	Azadeh et al., 2019; Le-Anh and De Koster, 2006; Lee et al., 2019; Qi et al., 2018

Fig. 7.1   Overview of managerial decision framework. (Source: Authors)

7.1   FURTHER CONSIDERATIONS ON CHANGE MANAGEMENT

Change, in general, is motivated by factors such as increased competition, global sales, sourcing strategies, financial markets, changes in business practices, shorter product life cycles, and rapid technological development. Change is a necessity for enterprises across many functional areas, and implementations of automation technology or mobile robots in a warehouse are certainly commonplace. The preceding sections of this book have concentrated on decisions around why to change to a new mobile robot system and the sequence of decisions that are likely to be involved from strategic, tactical, and operational angles. This section will focus on concepts and principles of change management with the aim to provide advanced guidance for decision-makers and their mobile robot implementations.

7.1.1   Incremental Change Versus Change Management

Organisational change is the alignment to internal or external forces or requirements, motivated not only by the need to align to competitive and technological challenges and requirements but also by the need to antici-pate those changes (Kerber and Buono 2005). A variety of approaches to managing change can be identified in the literature, broadly classified into systematic change methods and change management methods. Systematic change methods address incremental change for process adjustments, whilst change management methods are targeted at aligning change initia-tives to the organisational mission and strategy (Al-Haddad and Kotnour 2015; Worren et al. 1999). Systematic change methods comprise a set of procedures and tools developed over the past decades. These methods are incremental and have common elements, such as situation analysis, change planning, and communication and change implementation. Change man-agement methods are of wider applicability and have a broader approach targeting the company, including mission, organisational strategy, and employee involvement (Al-Haddad and Kotnour 2015).

Mobile robot implementations may be incremental improvements if only targeted at a single installation. However, the managerial decision framework for mobile robot implementation targets all aspects of change, strategic, tactical, and operational decisions as well as multiple applica-tions. The following subsection will therefore focus on the broader change management methods which are more applicable to this context of change management.

7.1.2   Change Management Methods

The literature identifies more than 30 different models of change (Bengat et al. 2015). Research about change management started around 1950 with the work of Lewin focusing on field theory, applied behavioural sci-ence, action research, and planned change. Lewin's still current approach of planned change is based on a three-step change model of unfreeze, act and move, and refreeze and has been modified or extended by other researchers. The following few subsections will focus on a selection of seminal change management methods, including Lewin (1947), Judson (1991), and Kotter (1996, as cited in Kotter and Schlesinger 2008). An overview of the methods is provided in Table 7.1 in relation to three key

Table 7.1 Change management methods

Author	Method	Phases			Focus
		Pre-phase	Transition	Post-phase	
Lewin (1947)	Lewin's method	1. Unfreeze	2. Act and move	3. Refreeze	– Linear approach – Stable environment – Incremental change
Judson (1991)	Judson's method	1. Analyse and plan change 2. Communicate the change 3. Gain acceptance of new behaviours	4. Change from status quo to a desired state	5. Consolidate and institutionalise the new state	– Predicting change barriers – Overcoming barriers to change – Communication and incentives for change
Kotter (1996)	Leading change	1. Establishing a sense of urgency 2. Creating the guiding coalition 3. Developing a vision and strategy	4. Communi cating the change vision 5. Empowering broad-based action 6. Generating short-term wins	7. Consolidating gains and producing more change 8. Anchoring new approaches in the culture	– People-driven change – Change in people's behaviour

Source: Authors

stages in change management, that is, the planning or pre-phase, the actual transition, and the post-phase, which anchors the change within the organisation.

Lewin's Method

In the pre-phase or unfreeze stage of Lewin's method (1947), the stable behaviour is destabilised, and the driving forces towards change are reinforced. This is a period of a break-up of the organisational complacency in order to prepare the solidified organisational structures. In step two, the transition phase, the organisation processes or structures will be led to a new equilibrium. This phase is mostly marked by uncertainty and confusion, whilst new procedures and thinking replace the previous behaviour. In a final step, the refreezing phase institutionalises the new behaviours and attitudes and stabilises the organisation with a different mindset on a new equilibrium (Ford 2009; Sarayreh et al. 2013).

Critics to Lewin's approach point out that change is often a non-linear and dynamic process which cannot be managed as a sequence of events (Ford 2009). Further, the approach was initially focused on smaller and incremental changes so that organisations can be moved from one stable state to another, with the prerequisite that the organisational environment is also stable. It should also be considered that the underlying assumption concerning cooperation and agreement of all stakeholders is not always given (Todnem By 2005).

Judson's Method
Judson's method of change comprises five steps. As outlined in Table 7.1, steps one to three align with Lewin's step one, whilst steps four and five are aligned with Lewin's steps two and three.

Judson sees the resistance to change as the most dominant barrier driven by the psychological and social dynamics that dominate people's behaviour. The resulting resistance impacts affected employees as well as managers that are essential for enabling the change through economic effects, personal security and convenience, job satisfaction, interpersonal relationships, satisfaction with management, and change in values (Kusmierek 2001). Judson's measures to overcome resistance include alternative media, reward programmes, and bargaining and persuasion (Armenakis and Bedeian 1999).

Kotter's Method
Kotter (1995, as cited in Kotter and Schlesinger 2008) introduced the leading change method. The key contribution of Kotter's method is his scope on the beginning of the change initiative with a careful planning and building of a solid foundation by a people-driven approach that enables the other steps.

Kotter's approach raises questions regarding the rigidity of the method, the irrelevance of some steps in specific contexts, and dealing with difficulties during a change management initiative. There is also the difficulty of studying change management projects in order to investigate if Kotter's method can be empirically verified (Sarayreh et al. 2013).

Saying that, Kotter's method offers a more complete approach that can be considered applicable to change management in warehouse automation as it aligns well with the various decisions outlined in the managerial decision framework for mobile robot implementation.

With regard to warehouse applications, a useful case study is provided by Moeller et al. (2016), which follows the implementation of warehouse automation within a central distribution centre of a global market leader for building materials. This case showcases a successful implementation that mirrors the leading change method by Kotter (1996, as cited in Kotter and Schlesinger 2008) as it involves the various steps outlined through the phases to a large extent. Mobile robot implementations are equally complex with regard to decisions, change requirements, stakeholder involvements, and change outcomes. Further, the required focus on processes, technology, and people points towards the leading change method as an applicable option.

7.1.3   Change Characteristics

Change can be differentiated into different change types characterised by the scope and depth of a change, change direction, as well as levels of control. Based on the changes required for implementing mobile robots in a warehouse environment, different characteristics are relevant. Hence, these change characteristics are further explored in the following subsections, with a summary of characteristics specific to mobile robot implementations provided in Table 7.2.

Change Scope and Depth
Scope of change refers to either evolutionary change or revolutionary change. Evolutionary or incremental change refers to small changes that modify small aspects within an organisation but do not change the overall working environment. Revolutionary, strategic, or transformational changes are more radical transformations, generally aimed at gaining a significant competitive advantage and affecting the essential capabilities of an organisation (Val and Fuentes 2003; Vilkas and Stancikas 2005).

Three depths of change are differentiated in literature. Type one comprises the operational base of an organisation's processes, procedures, and logistics, whilst type two changes target employees' performance. Both depths of change reshape tangible issues such as processes or performance management. Type three changes go beyond such operational boundaries, targeting organisational beliefs and values. It thus changes the norms and priorities assigned to organisational concepts and underlying ideas, actions, and outcomes (Bailey and Raelin 2015; Bartunek and Moch 1987).

Table 7.2   Change characteristics in mobile robot implementations

Change characteristics	Mobile robot implementation—change processes
Method	The leading change method provides the most detailed overview of activities with a focus on processes and people
Scope	Revolutionary for initial implementations
	Evolutionary for improvements and subsequent implementations
Depth	Process and technology level
	Social level
Direction	Top-down directed
	Bottom-up supported
Control	Planned change
	Applications/periods of directed and guided change
Frequency	Episodic change for the actual implementations
	The continuous change aimed at optimisation and further implementations

Source: Authors

In terms of change scope and depth, mobile robot implementations can be revolutionary to an organisation as they offer operational benefits that can lead to significant warehouse performance outcomes. Whilst initial or first implementations may indeed be considered truly strategic and transformational, successive implementations in additional facilities could be considered repetitive or evolutionary. In terms of depth, mobile robot implementations generally transverse process levels, employee, and performance management.

Change Direction
Change can be top-down coordinated or bottom-up. A top-down change is generally considered to be planned and transformational with a revolutionary and episodic character. On the other side, bottom-up change is an emergent and incremental process with an evolutionary character (Bouckenooghe 2010). An incremental approach can be applied to realise extended organisational innovations like learning new practices, the implementation of quality management, or impacts on the company culture. A top-down planned change approach is generally used for process innovations, although there must be room for incremental steps as well (Boer and During 2001).

Regarding the transformation of warehouse practices through new technology, it is generally given that the change is initiated through a top-down approach, that is, initially decided and funded by management. Within this kind of change, stakeholders from various warehouse operations are likely to be involved or impacted. Hence, certain parts of the change may be better facilitated through bottom-up support. This may include aspects such as employee involvement, training, and change communication—activities that are also reflected in the change management steps of the leading change method (Kotter 1996, as cited in Kotter and Schlesinger 2008).

Change Control

In terms of control, literature distinguishes between directed change, planned change, and guided change (Kerber and Buono 2005).

Directed change is a top-down approach that, facilitated through authority and compliance, focuses on emotional reactions to change. Leaders create and announce change and try to convince employees through business necessity, logical arguments, and emotional appeal. It is an authoritarian way of enabling change in an organisation and thus carries the risk of increasing resistance in the form of denial, anger, and unwillingness. However, when faced with limited change complexity and low socio-technological uncertainty, directed change may be a preferable choice.

As change complexity increases, a planned change is generally followed, akin to Lewin's three-stage process of unfreezing, changing, and refreezing. Planned change can start on every organisational layer instructed by top management. It is then supported by adequate support measures to soften impacts on productivity, to raise change acceptance, and to lower resistance.

The third approach—guided change—evolves from within the organisation. Main drivers are people's commitment and contributions to the purpose of the company. It thus focuses on further change development and follows as well as strengthens existing change processes within the organisation. Guided change follows the three-stage process mentioned but allows an organisation to identify organisational forces opposing change and ultimately to eliminate or even avoid such internal obstacles and blockages.

Mobile robot implementations are characterised by technical and business complexity on the one hand, as well as social and process uncertainty on the other hand. Whilst certain parts of the change process may be best organised as directed with clear messages and instructions, other parts of such a complex change need to be supported from within the organisation. Hence, change control is likely to depend on a planned change that involves directed managerial guidance combined with internal-oriented initiatives.

Change Frequency

The change management literature broadly distinguishes between episodic change and continuous change. Episodic changes are infrequent, discontinuous, and intentional with a more short-term focus on the adaptation of internal structures to new business requirements. Examples include technology implementations or changes in workforce. From a change management perspective, episodic changes follow a high-level process as outlined in Table 7.1 of unfreeze, transition, and refreeze. Episodic change is thus aligned with Lewin's planned change of transforming organisational characteristics by changing the status quo and transitioning the organisation to a new, stable work environment (Weick and Quinn 1999).

Evolving cumulative organisational alignments are known as continuous change and generally follow a cycle of freeze, transition or rebalance, and unfreeze. The assumption behind the continuous change is that the organisation is capable of improving work processes and social practices on a continuous basis. Further, such change may consist of continuous small adjustments and adaptions taking place across organisational units which, in combination, may be considered a substantial change. Interdependence between the organisational units helps to spread learnings and change processes from one unit to across the organisation. Within the freeze stage, the emphasis is on capturing and defining new processes. The rebalance phase comprises the reinterpretation or re-sequencing of work patterns for dissemination in the organisation. The unfreeze phase is to restart the improved processes, including the idea of continuous learning (Weick and Quinn 1999).

Mobile robots are a novel technology in warehouses, and the implementation can be characterised by resistance and inertia. Inertia towards getting such a technology shift started in the first place, and resistance before and during the change may be caused by existing management structures, work processes and routines, and business culture. Overcoming these and related factors necessitates a suitable change management approach. Initial technology changes within a limited number of ware-houses can be considered an episodic change. On the other side, continu-ous change would be aimed at the optimisation of existing mobile robot implementations, the dissemination of learnings and improved work pro-cesses across the organisation, and further new implementation projects.

7.2   FURTHER CONSIDERATIONS ON ALGORITHMS

The managerial decision framework needs to be complemented with a growing knowledge of algorithms and methods to manage robots in ware-houses, especially those serving chaotic contexts such as omni-channel and e-commerce. There is a lack of empirical work on the state-of-the-art prac-tices of managing mobile robot systems under realistic settings. As the number of companies implementing such solutions increases, the oppor-tunities for conducting applied research on operational decisions of the managerial framework will also expand.

Many algorithms have been proposed to manage mobile robot opera-tions within the warehouse. Each has varying performance under a range of operating scenarios. One of the typical problems encountered in a ware-house with mobile robots is congestion. As the material handling volumes increase, the number of mobile robots in circulation also increases, leading to congestion. Capacitated vehicle routing problems are extended to warehouse settings with modifications to incorporate conflict-free routing and storage allocation (Thanos et al. 2019). Owing to the complexity of vehicle routing problems, heuristics are proposed to reach a feasible solu-tion within an acceptable time frame. A promising line of research includes extending current models to accommodate dynamic updates to the system as same-day delivery services become more prevalent.

With any new system of scheduling, the expected improvement to the system informs the decision to deploy it. Yoshitake et al. (2019) design a scheduling system for automated guided vehicles for transportation of inventory shelves and sorting shelves. Inventory shelves are where the products are stored and should be picked from, whereas sorting shelves are those that include shipping boxes and picked items are put into these sorting shelves for dispatch. The novelty of their approach is to introduce the movement of sorting shelves, which are otherwise fixed until comple-tion in conventional warehouse systems. Whether to deploy the schedul-ing algorithm depends on the expected productivity increase, which is reported to be around 6.7% for normal and 12.5% in actual warehouses.

Such productivity metrics depend on the number of picking items, total time of waiting, and total time of picking where the conventional system is compared with the proposed scheduling system.

Wang et al. (2020) propose a design framework for deploying mobile robots in warehouses which presents the trade-offs between system performance and costs, allowing the decision-maker to choose a design-based system throughput and average flow time as key performance indicators. The design framework allows rapid assessment of a range of designs that differ by the number of modules, number of aisles and columns, and the layout of each module. In this context, a module comprises a workstation, a picker, and robots in a pick area with aisles dedicated to the robots. Such experimental designs support warehouse managers to find the most appropriate layout and system design to meet customer requirements.

The width of the aisles in a warehouse significantly affects the operational decision rules of mobile robots. For example, in some warehouses, aisles are too narrow that two robots cannot pass each other, in either direction. Such a constraint necessitates sequencing of the aisle access. Polten and Emde (2020) develop a mixed-integer programming model to sequence access and solve the model using a neighbourhood search. This specific problem is named multi-aisle access scheduling problem, which assigns a given set of storage or retrieval jobs to a set of mobile robots, resulting in the sequence of jobs to be performed by robots. The model has several assumptions found in other scheduling problems, such as once a job has started, it cannot be interrupted until it is complete. Polten and Emde (2020) develop a large neighbourhood search heuristic which can handle several different versions of the problem and outperforms CPLEX in terms of proximity to the lower bound for the problem. The authors' work has significant managerial implications. If parallel access policy is used, there is little advantage of wide aisles over narrow aisles because the advantage of having no blockage inside the aisle is offset by longer travel times. Fewer robots than the number of aisles are recommended with parallel access policy, as the policy increases the utilisation of the robots.

In many warehouse environments, dynamically updating the robot's navigation path is required as multiple entities change locations and positions on a continuous basis. Algorithms such as A* and D* are developed with different objectives. The A* algorithm finds the shortest path from origin to destination (Ng et al. 2020), whereas the D* algorithm devises a path from the destination to the source (current location of the robot) under unknown, partially known, or changing environments, updating each time the robot encounters an obstacle (Stentz 1993). Ng et al. (2020) develop a novel algorithm that they name AD*, which can detect static and dynamic obstacles. The robot processes the image of the warehouse environment, identifies obstacles on its path, and predicts the trajectory of such obstacles.

Similar to dynamic obstacle avoidance, another relevant line of research is centred on incorporating dynamic demand information into mobile robot scheduling. The need has arisen mainly owing to the increasingly competitive e-commerce market where retailers have introduced same-day deliveries and expedited deliveries. Li et al. (2020) develop a mathematical model for scheduling robots operating in warehouses facing dynamic customer demand. They use particle swarm optimisation to solve the task allocation problem and show via simulations that particle swarm optimisation outperforms state-of-the-art genetic algorithms. The authors recognise that their work ignores collision avoidance, which is a significant issue for real operating environments. Another area for extending their work is to include robot break-downs or stoppages due to recharging explicitly into the modelling exercise.

Complementary to Li et al. (2020), Lee et al. (2020) propose an efficient genetic algorithm to schedule and route multiple robots. Their approach can design schedules for thousands of routes and integrate them on a daily basis. While an optimal solution is computationally too expensive, their combinatorial auction-based genetic algorithm achieves close-to-optimal results in comparably shorter run times (gaps of 2–12% in less than 1 minute run time).

Several works presented in detail in this section are conceptual and theoretical, working on hypothetical or simulated data to establish the performance of the proposed algorithms. There are open questions on how the algorithms may perform under real operating conditions and how they would cope with the occasional seasonality observed in warehouses on special retail days such as Black Friday, Cyber Monday, Christmas, Easter, or retailer-organised promotion periods.

Companies should review their operational decisions regularly to respond to changing trends in purchasing patterns both in terms of activities needed (e.g. number of items per order directly affects the path planning and picking) and requirements (package size and characteristics based on products shipped). In addition to robot type and capacity, the scalability of the entire operation should be considered. With the advancement of the Internet of Things and big data analytics, it is becoming feasible to incorporate real-time or near-real-time data into the execution of algorithms. One question pending for decision-makers is the return on investment, which is highly dependent on the operational characteristics of the warehouse where the mobile robot systems will be implemented.

REFERENCES

Al-Haddad, S. and Kotnour, T. (2015) 'Integrating the organizational change literature: a model for successful change', Journal of Organizational Change Management, 28(2) Bradford: Emerald Group Publishing, Limited, p. 234.

Armenakis, A.A. and Bedeian, A.G. (1999) 'Organizational Change: A Review of Theory and Research in the 1990s', Journal of Management, 25(3), pp. 293–315.

Azadeh, K., De Koster, R. B. M., & Roy, D. (2019). Robotized and automated warehouse systems: Review and recent developments. In Transportation Science (Vol. 53, Issue 4). https://doi.org/10.1287/trsc.2018.0873

Bailey, J.R. and Raelin, J.D. (2015) 'Organizations Don't Resist Change, People Do: Modeling Individual Reactions to Organizational Change Through Loss and Terror Management.', Organization Management Journal (Routledge), 12(3), pp. 125–138.

Bartunek, J.M. and Moch, M.K. (1987) 'First-Order, Second-Order, and Third-Order Change and Organization Development Interventions: A Cognitive Approach', The Journal of Applied Behavioral Science, 23(4), pp. 483–500.

Bechtsis, D., Tsolakis, N., Vlachos, D., & Iakovou, E. (2017). Sustainable supply chain management in the digitalisation era: The impact of Automated Guided Vehicles. Journal of Cleaner Production, 142, 3970–3984. https://doi.org/10.1016/j.jclepro.2016.10.057

Bengat, J., Odenyo, M., & Rotich, J. (2015). Organizational change and resistance dilemmas resolution approaches and mechanisms. International Journal of Economics, Commerce and Management, 3(2), 1-16.

Boer, H. and During, W.E. (2001) 'Innovation, what innovation? A comparison between product, process and organisational innovation', International Journal of Technology Management, 22(1/2/3), p. 83.

Bouckenooghe, D. (2010) 'Positioning Change Recipients' Attitudes Toward Change in the Organizational Change Literature', Journal Of Applied Behavioral Science, 46(4) Arlington: Sage Publications, Inc., pp. 500–531.

Boysen, N., De Koster, R. B. M., & Weidinger, F. (2019). Warehousing in the e-commerce era: A survey. In European Journal of Operational Research (Vol. 277, Issue 2, pp. 396–411). Elsevier B.V. https://doi.org/10.1016/j.ejor.2018.08.023

Claes, D., Oliehoek, F., Baier, H., & Tuyls, K. (2017). Decentralised Online Planning for Multi-Robot Warehouse Commissioning.

Dou, J., Chen, C., & Yang, P. (2015). Genetic Scheduling and Reinforcement Learning in Multirobot Systems for Intelligent Warehouses. Mathematical Problems in Engineering, 2015. https://doi.org/10.1155/2015/597956

Draganjac, I., Miklic, D., Kovacic, Z., Vasiljevic, G., & Bogdan, S. (2016). Decentralized Control of Multi-AGV Systems in Autonomous Warehousing Applications. IEEE Transactions on Automation Science and Engineering, 13(4), 1433–1447. https://doi.org/10.1109/TASE.2016.2603781

Fazlollahtabar, H., & Saidi-Mehrabad, M. (2013). Methodologies to Optimize Automated Guided Vehicle Scheduling and Routing Problems: A Review Study. Journal of Intelligent and Robotic Systems: Theory and Applications, 77(3–4), 525–545. https://doi.org/10.1007/s10846-013-0003-8

Ferrara, A., Gebennini, E., & Grassi, A. (2014). Fleet sizing of laser guided vehicles and pallet shuttles in automated warehouses. International Journal of Production Economics, 157(1), 7–14. https://doi.org/10.1016/j.ijpe.2014.06.008

Ford, M.W. (2009) 'Size, structure and change implementation', Management Research News, 32(4) Department of Management, College of Business, Northern Kentucky University, Highland Heights, KY, United States, pp. 303–320.

Hamann, H., Markarian, C., Auf Der Heide, F. M., & Wahby, M. (2018). Pick, Pack, & Survive: Charging Robots in a Modern Warehouse based on Online Connected Dominating Sets. Leibniz International Proceedings in Informatics, LIPIcs, 100, 221–2213. https://doi.org/10.4230/LIPIcs.FUN.2018.22

Huang, G. Q., Chen, M. Z. Q., & Pan, J. (2015). Robotics in ecommerce logistics. HKIE Transactions Hong Kong Institution of Engineers, 22(2), 68–77. https://doi.org/10.1080/1023697X.2015.1043960

Inam, R., & Raizer, K. (2018). Risk Assessment for Human-Robot Collaboration in an automated warehouse scenario.

Judson, A.S. (1991) Changing behaviour in organizations: minimizing resistance to change. Cambridge, MA.: Blackwell.

Kattepur, A., Rath, H. K., Simha, A., & Mukherjee, A. (2018). Distributed optimization in multi-agent robotics for industry 4.0 warehouses. Proceedings of the ACM Symposium on Applied Computing, 808–815. https://doi.org/10.1145/3167132.3167221

Kerber, K.W. and Buono, A.F. (2005) 'Rethinking organizational change: Reframing the challenge of change management', Organization Development Journal, 23(3) Bentley College, Simmons College, United States, pp. 23–38.

Kotter, J. P. (1995). The New Rules: How to Succeed in Today's Post-corporate World. New York: Free Press.

Kotter, J. P. (1996). Leading Change. Boston: Harvard Business School Press.

Kotter, J.P. and Schlesinger, L.A. (2008) 'Choosing Strategies for Change', Harvard Business Review, July/ August, pp. 130–139.

Krnjak, A., Draganjac, I., Bogdan, S., Petrovic, T., Miklic, D., & Kovacic, Z. (2015). Decentralized control of free ranging AGVs in warehouse environments. In Proceedings - IEEE International Conference on Robotics and Automation (Vols. 2015-June, Issue June). https://doi.org/10.1109/ICRA.2015.7139465

Kusmierek, K.N. (2001) Understanding and Addressing Resistance to Organizational Change. Ann Arbor.

Lamballais, T., Roy, D., & De Koster, R. B. M. (2017). Estimating performance in a Robotic Mobile Fulfillment System. European Journal of Operational Research, 256(3), 976–990. https://doi.org/10.1016/j.ejor.2016.06.063

Lamballais, T., Roy, D., & De Koster, R. B. M. (2020). Inventory allocation in robotic mobile fulfillment systems. IISE Transactions, 52(1), 1–17. https://doi.org/10.1080/24725854.2018.1560517

Le-Anh, T., & De Koster, R. B. M. (2006). A review of design and control of automated guided vehicle systems. European Journal of Operational Research, 171(1), 1–23. https://doi.org/10.1016/j.ejor.2005.01.036

Le-Anh, T., De Koster, R. B. M., & Yu, Y. (2010). Performance evaluation of dynamic scheduling approaches in vehicle-based internal transport systems. International Journal of Production Research, 48(24), 7219–7242. https://doi.org/10.1080/00207540903443279

Lee, C. W., Wong, W. P., Ignatius, J., Rahman, A., & Tseng, M. L. (2020). Winner determination problem in multiple automated guided vehicle considering cost and flexibility. Computers and Industrial Engineering, 142. https://doi.org/10.1016/j.cie.2020.106337

Lee, C., Lin, B., Ng, K. K. H., Lv, Y., & Tai, W. C. (2019). Smart robotic mobile fulfillment system with dynamic conflict-free strategies considering cyber-physical integration. Advanced Engineering Informatics, 42. https://doi.org/10.1016/j.aei.2019.100998

Lewin, K. (1947) 'Frontiers in Group Dynamics', Human Relations, 1(1) SAGE Publications, pp. 5–41.

Li, Z., Barenji, A. V., Jiang, J., Zhong, R. Y., & Xu, G. (2020). A mechanism for scheduling multi robot intelligent warehouse system face with dynamic demand. Journal of Intelligent Manufacturing, 31(2), 469–480. https://doi.org/10.1007/s10845-018-1459-y

Lienert, T., Stigler, L., & Fottner, J. (2019). Failure-Handling Strategies For Mobile Robots In Automated Warehouses. ECMS 2019 Proceedings Edited by Mauro Iacono, Francesco Palmieri, Marco Gribaudo, Massimo Ficco, 199–205. https://doi.org/10.7148/2019-0199

Merschformann, M., Lamballais, T., De Koster, R. B. M., & Suhl, L. (2019). Decision rules for robotic mobile fulfillment systems. Operations Research Perspectives, 6. https://doi.org/10.1016/j.orp.2019.100128

Moeller, K., Gabel, J., & Bertagnolli, F. (2016). Fischer Fixing Systems: Moving Forward With The Workforce - Change Communication At The Global Distribution Center. Journal of Information Technology Education: Discussion Cases, 5(September 2017), 01. https://doi.org/10.28945/3457

Ng, M. K., Chong, Y. W., Ko, K. man, Park, Y. H., & Leau, Y. B. (2020). Adaptive path finding algorithm in dynamic environment for warehouse robot. Neural Computing and Applications. https://doi.org/10.1007/s00521-020-04764-3

Polten, L., & Emde, S. (2020). Scheduling automated guided vehicles in very narrow aisle warehouses. Omega (United Kingdom). https://doi.org/10.1016/j.omega.2020.102204

Qi, M., Li, X., Yan, X., & Zhang, C. (2018). On the evaluation of AGVS-based warehouse operation performance. Simulation Modelling Practice and Theory, 87, 379–394. https://doi.org/10.1016/j.simpat.2018.07.015

Roy, D., Nigam, S., De Koster, R. B. M., Adan, I., & Resing, J. (2019). Robot-storage zone assignment strategies in mobile fulfillment systems. Transportation Research Part E: Logistics and Transportation Review, 122, 119–142. https://doi.org/10.1016/j.tre.2018.11.005

Sarayreh, B.H. et al. (2013) 'Comparative Study: The Kurt Lewin of Change Management', International Journal of Computer and Information Technology, 2(4), pp. 626–629.

Sartoretti, G., Kerr, J., Shi, Y., Wagner, G., Satish Kumar, T. K., Koenig, S., & Choset, H. (2019). PRIMAL: Pathfinding via Reinforcement and Imitation Multi-Agent Learning. In IEEE Robotics and Automation Letters (Vol. 4, Issue 3). https://doi.org/10.1109/LRA.2019.2903261

Schmidt, T., & Schulze, F. (2009). Future approaches to meet complexity requirements in material handling systems. FME Transactions, 37(4), 159–166.

Stentz, A. (1993). Optimal and efficient path planning for unknown and dynamic environments. Carnegie-Mellon Univ Pittsburgh Pa Robotics Inst.

Thanos, E., Wauters, T., & Vanden Berghe, G. (2019). Dispatch and conflict-free routing of capacitated vehicles with storage stack allocation. Journal of the Operational Research Society. https://doi.org/10.1080/0160568 2.2019.1595191

Todnem By, R. (2005) 'Organisational change management: A critical review', Journal of Change Management, 5(4), pp. 369–380.

Val, M.P. and Fuentes, C.M. (2003) 'Resistance to change: a literature review and empirical study', Management Decision, 41(2), pp. 148–155.

Vilkas, M. and Stancikas, E.R. (2005) 'Typology of organization's processes', Inzineriné Ekonomika, 3(43) Kaunas University of Technology, pp. 35–41.

Vis, I. F. A. (2006). Survey of research in the design and control of automated guided vehicle systems. European Journal of Operational Research, 170(3), 677–709. https://doi.org/10.1016/j.ejor.2004.09.020

Vivaldini, K., Rocha, L. F., Becker, M., & Moreira, A. P. (2015). Comprehensive review of the dispatching, scheduling and routing of AGVs. Lecture Notes in Electrical Engineering, 321 LNEE, 505–514. https://doi.org/10.1007/978-3-319-10380-8_48

Wang, W., Wu, Y., Zheng, J., & Chi, C. (2020). A comprehensive framework for the design of modular robotic mobile fulfillment systems. IEEE Access, 8, 13259–13269. https://doi.org/10.1109/ACCESS.2020.2966403

Weick, K.E. and Quinn, R.E. (1999) 'Organizational Change And Development', Annual Review of Psychology, 50(1), pp. 361–386.

Witczak, M., Majdzik, P., Stetter, R., & Lipiec, B. (2020). A fault-tolerant control strategy for multiple automated guided vehicles. Journal of Manufacturing Systems, 55, 56–68. https://doi.org/10.1016/j.jmsy.2020.02.009

Worren, N.A.M. et al. (1999) 'From Organizational Development to Change Management', The Journal of Applied Behavioral Science, 35(3), pp. 273–286.

Xu, J., Wang, J., & Chen, W. (2019). An efficient recharging task planning method for multi-robot autonomous recharging problem. In IEEE International Conference on Robotics and Biomimetics, ROBIO 2019. https://doi.org/10.1109/ROBIO49542.2019.8961439

Yan, R. D., Dunnett, S. J., & Jackson, L. M. (2018). Optimising the maintenance strategy for a multi-AGV system using genetic algorithms. In Safety and Reliability – Safe Societies in a Changing World. CRC Press. https://doi.org/10.1201/9781351174664

Yan, R., Jackson, L. M., & Dunnett, S. J. (2017). Automated guided vehicle mission reliability modelling using a combined fault tree and Petri net approach. International Journal of Advanced Manufacturing Technology, 92(5–8), 1825–1837. https://doi.org/10.1007/s00170-017-0175-7

Yoshitake, H., Kamoshida, R., & Nagashima, Y. (2019). New Automated Guided Vehicle System Using Real-Time Holonic Scheduling for Warehouse Picking. IEEE Robotics and Automation Letters, 4(2), 1045–1052. https://doi.org/10.1109/LRA.2019.2894001

Yuan, Z., & Gong, Y. Y. (2017). Bot-in-time delivery for robotic mobile fulfillment systems. IEEE Transactions on Engineering Management, 64(1), 83–93. https://doi.org/10.1109/TEM.2016.2634540

Zavadskas, E. K., Nunić, Z., Stjepanović, Ž., & Prentkovskis, O. (2018). Novel Rough Range of Value Method (R-ROV) for selecting automatically guided vehicles (AGVs). Studies in Informatics and Control, 27(4), 385–394. https://doi.org/10.24846/v27i4y201802

Zou, B., Xu, X., Gong, Y. (Yale), & De Koster, R. B. M. (2018). Evaluating battery charging and swapping strategies in a robotic mobile fulfillment system. European Journal of Operational Research, 267(2), 733–753. https://doi.org/10.1016/j.ejor.2017.12.008

Zou, Y., Zhang, D., & Qi, M. (2019). Order picking system optimization based on picker-robot collaboration. ACM International Conference Proceeding Series, 1–6. https://doi.org/10.1145/3364335.3364386

CHAPTER 8

Research Agenda

This chapter discusses the mobile robot system selection according to Table 3.3 and the managerial decision framework presented in Table 7.1. At the end of each subsection, research directions are outlined which, in combination, form a research agenda for mobile robot systems in warehouses.

8.1 MOBILE ROBOT SYSTEMS AND SELECTION CRITERIA

This book mentions five criteria for evaluating and choosing between different mobile robot systems (Chap. 3). The criteria outlined are derived from and supported by literature, whilst the authors specifically developed the rating system to support decision-makers in warehouses. The rationale of this selection support tool for mobile robot systems and the ratings of each system were illustrated with regard to their applicability in supply chain practice. In combination, this forms a key contribution of this study, which could be further supported by the following research directions.

- A list of criteria to select the correct mobile robot system (i.e. cost, flexibility in infrastructure, flexibility in material handling, scalability, time-to-implement) could be developed and refined through practical applications with empirical evidence to aid decision-makers in choosing the best solution.

A. Yildirim et al., *Mobile Robot Automation in Warehouses*, Palgrave Studies in Logistics and Supply Chain Management, https://doi.org/10.1007/978-3-031-12307-8_8

- More than half of the empirical papers focus on picking operations. Even though it is the most time-consuming and costliest activity in the warehouse, other operations such as sortation, put-away, and loading/unloading should also be evaluated when considering mobile robot system implementation.
- The performance (e.g. throughput, labour productivity) of mobile robot systems other than barcode-guided mobile robots has not been systematically evaluated due to a lack of simulations and real-life applications. Theory and practice should work together to define relevant performance indicators in this regard and subsequently analyse performance areas considering different user scenarios.

8.2   MANAGERIAL DECISION FRAMEWORK

The hierarchical decision framework (Table 7.1) is targeted at the design, planning, and management of warehouses adopting mobile robot technologies in the digitalisation era. It synthesises decision areas that have been identified through the systematic literature review and organises these across managerial decision horizons. It considers the various choices put forward by the reviewed papers in order to capture a wide array of decision areas and criteria. Research areas stemming from the framework as a whole are as follows.

- To test and/or improve this conceptual framework's structure and robustness with more focus areas and/or decision questions, a further investigation needs to be carried out with experts and practitioners familiar with mobile robot implementation.
- The framework should be put into practice with various mobile robot systems to illuminate the differences within the focus areas in distinct systems.

The following subsections provide further reflection and provide research directions aimed at enhancing the applicability and generalisability of the managerial decision framework.

8.2.1   Strategic-Level Focus Areas

This book limits the strategic level to pre-implementation decisions, which would require a considerable amount of time (maybe a year) and investment to implement selected technologies. For instance, Bechtsis et al. (2017) consider the number of robots (fleet sizing) as a strategic decision, whereas Le-Anh and De Koster (2006) and Wang et al. (2020) suggest that it is of tactical nature. Although the initial number of robots is decided before the implementation of a solution, the total number may obviously be altered in the medium term, at the tactical level. Keeping in mind this logic regarding strategic-level decisions, the following pre-implementation research areas are identified.

- More than half of the studies do not mention how robots would be coordinated in the warehouse. To identify the applicability of robot coordination types in different scenarios, it should be explicitly considered in empirical studies.
- Most papers assume a fixed layout which does not adequately reflect the flexible use of mobile robots and decreases the generalisability of the results. Thus, researchers should incorporate alternative layouts into the implementation of mobile robot automation.
- Human-robot interaction is not mentioned adequately, and studying a mobile robot-only warehouse lacks practicality. Subjects such as 'change management' and 'human safety' require researchers' attention in order to support system implementations and raise managers' awareness of and sensibility to issues in human-robot interaction.

8.2.2   Tactical-Level Focus Areas

Tactical-level decisions may have impacts across time horizons and decision levels and, depending on the situation, will impact strategic or operational decisions to different degrees. For example, storage assignment is considered at the strategic level and the operational level in separate studies (Füßler et al. 2019). Further, the order management plan is a common subject, but none of the papers reviewed allocates it to a particular level. We consider it at the tactical level as it does not necessarily need to be decided before the implementation of the system, but observing its effect on the system would be in the medium term. In support of managerial decisions at a tactical level, the following research avenues are put forward.

- Storage assignment decision is one of the least attended subjects as it might be perceived as an inventory management decision rather than a mobile robot-related decision. However, decision-makers require support in this regard since considering how SKUs should be distributed in a warehouse holds the potential to decrease the number of robots needed and increase warehouse throughput.

- Accounting for the biggest portion of the variable cost, fleet sizing is a significant entry barrier for implementing mobile robot systems. Yet, many papers leave fleet size optimisation unattended as their main aim is to prove the feasibility of algorithms for mobile robot systems. Optimum fleet size considerations should thus be a focal point of research and should be included in empirical scenarios as well as in simulation exercises.

- Maintenance and failure handling strategies require research attention since disregarding robot failures can lead to unforeseen and undesirable delays in, for example, inventory movements or picking tasks. Hence, maintenance strategies should be evaluated and compared with regard to their inherent trade-offs regarding costs, frequency, time requirements, prevention effectiveness, and so on. Predictive maintenance strategy for robot fleets is a promising research direction which could be coupled with investigations into the success of specific actions in case of robot failures to maintain the continuity of the operations.

- Robot energy management is understudied despite its impacts on warehouse throughput, traffic congestion, fleet size considerations, and space requirements for charging stations. Empirical studies could help to evaluate the practical implications of energy management decisions and new technology developments, whereas modelling research could focus on optimisation approaches of different energy management strategies and their trade-offs.

8.2.3   Operational-Level Focus Areas

Operational-level decisions may be altered in the short term, and the results could be observed within the same week or day, such as task allocation of mobile robots. Even though they are mentioned as a tactical level decision by Le-Anh and De Koster (2006), most authors consider them at the operational level because they can be altered daily with a new algorithm.

- Dynamic task allocation approaches need to be adapted to the application of mobile robot systems as they hold the potential to improve managing chaotic warehouse environments where there is an increased chance of occurrences of unforeseen events.

- In path planning, instead of focusing on algorithms such as Dijkstra's and A* that provide optimal solutions, computationally scalable sub-optimal approaches should be studied for large warehouses.
- To decrease task completion times and avoid delays, studies should concentrate on proactive conflict and deadlock management instead of reactive approaches (Lee et al. 2019). Proactive approaches are scarce in the current literature even though they eliminate the time loss of two robots coming across each other.
- Even though many warehouses require large fleets of mobile robots in practice, academic studies often avoid this reality and somewhat simplify the scale of problems. Thus, suggested algorithms tend to be developed for unrealistic scenarios and should hence be tested for larger-scale applications of mobile robots. Additionally, their traffic management should be studied as the required level of sophistication and flexibility of a solution may increase at scale.

8.3   Research Agenda Summary

Figure 8.1 presents the research agenda of key topics for warehouse mobile robot systems. Each topic highlights an information deficiency or a necessity for elaboration on a decision-level-related subject. These subjects are a synthesis of the authors' observations and the research suggestions gathered from the reviewed papers.

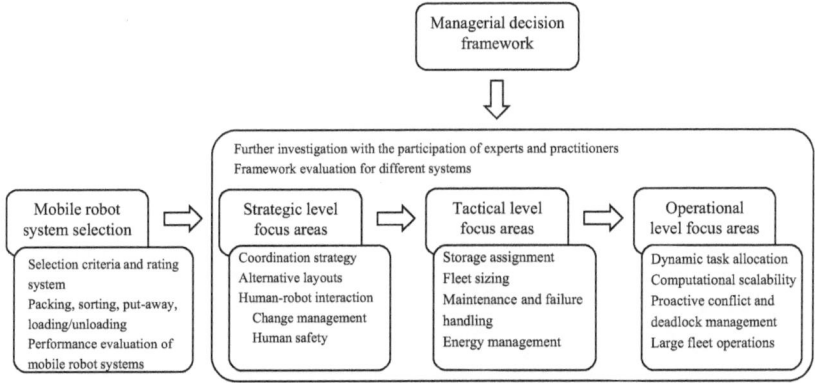

Fig. 8.1   Key topics for the research agenda. (Source: Authors)

REFERENCES

Bechtsis, D., Tsolakis, N., Vlachos, D., & Iakovou, E. (2017). Sustainable supply chain management in the digitalisation era: The impact of Automated Guided Vehicles. *Journal of Cleaner Production*, *142*, 3970–3984. https://doi.org/10.1016/j.jclepro.2016.10.057

Füßler, D., Boysen, N., & Stephan, K. (2019). Trolley line picking: storage assignment and order sequencing to increase picking performance. *OR Spectrum*, *41*(4), 1087–1121. https://doi.org/10.1007/s00291-019-00566-9

Le-Anh, T., & De Koster, R. B. M. (2006). A review of design and control of automated guided vehicle systems. *European Journal of Operational Research*, *171*(1), 1–23. https://doi.org/10.1016/j.ejor.2005.01.036

Lee, C., Lin, B., Ng, K. K. H., Lv, Y., & Tai, W. C. (2019). Smart robotic mobile fulfillment system with dynamic conflict-free strategies considering cyber-physical integration. *Advanced Engineering Informatics*, *42*. https://doi.org/10.1016/j.aei.2019.100998

Wang, W., Wu, Y., Zheng, J., & Chi, C. (2020). A comprehensive framework for the design of modular robotic mobile fulfillment systems. *IEEE Access*, *8*, 13259–13269. https://doi.org/10.1109/ACCESS.2020.2966403

CHAPTER 9

Conclusion

This book identified ten mobile robot systems used in warehouses through a systematic literature review of 130 papers from four databases. Based on the insights gained, a conceptual managerial decision framework was developed with thirteen strategic-, tactical-, and operational-level focus areas that require consideration before or during the application of mobile robot systems in warehouses. System evaluations through a criteria rating system and the framework combined will aid decision-makers in implementing a mobile robot solution step-by-step. The research avenues stemming from these artefacts will help academics illuminate the unattended topics, forming a balanced and complete guide to practice.

According to Llopis-Albert et al. (2019), managers pay more attention to management and financial issues (which are mainly strategic and tactical focus areas in our framework) rather than technical (which are primarily operational focus areas in our framework) issues. In contrast, academic papers in our review mainly focus on operational decisions rather than strategic and tactical decisions. There should be cooperation between the practice and the theory, and a balanced approach among strategic, tactical, and operational focus areas.

The authors believe that mobile robot automation is at an early stage. The sooner the firms consider implementing these solutions, the more benefit they will get by gathering experience. For the warehouses with minor or no automation, it is suggested to apply systems for the basic

A. Yildirim et al., *Mobile Robot Automation in Warehouses*, Palgrave Studies in Logistics and Supply Chain Management, https://doi.org/10.1007/978-3-031-12307-8_9

warehouse operations. Further, instead of automating the whole operation, it is recommended to automate it up to a percentage. The reason is that there will be many unforeseen issues, such as the surface quality, shelf quality, and conflicts due to traffic, which even the technology provider cannot foresee before becoming operational.

An example can be implementing pallet-carrying mobile robot solutions to carry the pallets from the packing area to the outbound area and letting human-driven forklifts in the area to cover the potential low performance of mobile robots. This way, the personnel and the decision-makers would gather experience and become familiar with implementing such solutions. Then, more comprehensive solutions could be considered for better outcomes.

As with any study, there are limitations to this analysis. Firstly, the research reviewed is based on applicable academic databases only, focusing on studies from 2000 onwards. Thus, its outcomes might not completely reflect available mobile robot systems and all potential managerial focus areas of mobile robot applications in warehouses. Secondly, the managerial decision framework is conceptual, and its building blocks are exclusively derived from the identified papers. Hence, opportunities will be explored to test and develop the framework, incorporating implementation practices in future studies.

Reference

Llopis-Albert, C., Rubio, F., & Valero, F. (2019). Fuzzy-set qualitative comparative analysis applied to the design of a network flow of automated guided vehicles for improving business productivity. *Journal of Business Research, 101*, 737–742. https://doi.org/10.1016/j.jbusres.2018.12.076

APPENDIX: SYNTHESIS TABLE

129

Reference	Year	Type of Study	System Type	Warehouse Operation	Criteria for Mobile Robot System	Identifying KPIs	Robot Type	Robot Coordination	Facility Layout	Human-Robot Interaction	Storage Assignment Plan	Order Management Plan	Quantity of Robots (Fleet Sizing)	Maintenance and Failure Handling	Energy Management Plan	Robot Task-Allocation	Path Planning of Robots	Deadlock Resolution and Conflict Avoidance
De Koster et al., 2004	2004	Empirical	Not mentioned	Not mentioned	Not mentioned	Robot utilisation, task delay/response time	Not mentioned	Centralised	Not mentioned	Not mentioned	Not mentioned	Online/Dynamic	Not mentioned	Not mentioned	Not mentioned	Compare rules	Not mentioned	Not mentioned
Le-Koster & De Koster, 2006	2006	Review	Not mentioned	Not mentioned	Not mentioned	Robot travel time, travel distance, utilisation, number of deadlocks/conflicts/routing failures, task delay/response time	Not mentioned	Defines coordination types	Reviews flowpath of robots and strategies for idle robot positioning	Not mentioned	Not mentioned	Not mentioned	Reviews fleet sizing methods	Not mentioned	Reviews battery management	Reviews online/fixed task allocation	Reviews dynamic and static algorithms	Reviews zoning strategies, conflict management, and system workload balancing
Vis, 2006	2006	Review	Not mentioned	Not mentioned	Not mentioned	Not mentioned	Not mentioned	Not mentioned	Reviews flowpath of robots and rules for idle robot positioning	Not mentioned	Not mentioned	Not mentioned	Not mentioned	Not mentioned	Reviews battery management	Reviews rules	Reviews dynamic and static algorithms	Reviews conflict avoidance through flow-path of robots
Schulze & Zhao, 2007	2007	Review	Review wire using and laser-guided mobile robots	Not mentioned	Not mentioned	Not mentioned	AGV	Not mentioned	Not mentioned	Not mentioned	Not mentioned	Not mentioned	Not mentioned	Not mentioned	Not mentioned	Not mentioned	Not mentioned	Not mentioned
Lau et al., 2007	2007	Empirical	Not mentioned	Not mentioned	Not mentioned	Not mentioned	AMR	Distributed	Not mentioned	Not mentioned	Not mentioned	Not mentioned	Compares different fleet sizes	Migration of task in case of failure	Not mentioned	Artificial immune system based exploration	Not mentioned	Not mentioned
Wurman et al., 2008	2008	Empirical	Barcode-Guided Mobile Robots	Picking	Not mentioned	Not mentioned	AMR	Distributed	Not mentioned	Not mentioned	Not mentioned	Online/Dynamic	Compares different fleet sizes	Not mentioned	Not mentioned	Not mentioned	A*	Not mentioned
Schmidt & Schultze, 2009	2009	Review	Not mentioned	Not mentioned	Cost, service quality, material handling flexibility, infrastructure flexibility, and scalability	Not mentioned	Not mentioned	Not mentioned	Studies U and I layouts	Not mentioned	Not mentioned	Not mentioned	Not mentioned	Not mentioned	Not mentioned	Not mentioned	Not mentioned	Not mentioned
Le-Ahn et al., 2010	2010	Empirical	Not mentioned	Not mentioned	Not mentioned	Robot utilisation/idleness, task delay/response time	Not mentioned	Not mentioned	Not mentioned	Not mentioned	Not mentioned	Not mentioned	Not mentioned	Not mentioned	Not mentioned	Compare heuristics and rules	Shortest path algorithms such as Dijkstra's	Not mentioned

Study	Year	Type	Guidance technology	Process	Performance metric	Robot type	Control architecture	Warehouse scope	Fleet / optimisation	Batching	Fault tolerance	Battery	Task dispatching	Path planning	Traffic management
Samdi-Recak et al., 2010	2010	Empirical	Laser-Guided Mobile Robots	Not mentioned	Not mentioned	Not mentioned	Not mentioned	Not mentioned	Not mentioned	Not mentioned	Not mentioned	Not mentioned	Not mentioned	Shortest path algorithms such as Dijkstra's	Time windows based dynamic routing
Enright & Wurman, 2011	2011	Empirical	Barcode-Guided Mobile Robots	Picking	Not mentioned	AMR	Not mentioned	Not mentioned	Not mentioned	Not mentioned	Not mentioned	Not mentioned	Not mentioned	Not mentioned	Limits number of vehicles in zones and increases expected time for congested zones
Confessore et al., 2013	2013	Empirical	Not mentioned	Not mentioned	Avg task completion time	Not mentioned	Centralised	Not mentioned	Optimised number of robots as a conclusion	Not mentioned	Migration of task in case of failure	Charges when battery is low with deterministic time	Network flow based vehicle initiated dispatching	Not mentioned	Not mentioned
Faisal et al., 2013	2013	Empirical	Not mentioned	Not mentioned	Robot utilisation / idleness	AMR	Decentralised (Distributed)	Not mentioned	Not mentioned	Not mentioned	Not mentioned	Not mentioned	Not mentioned	Tracking fuzzy logic controller	Obstacles avoiding fuzzy logic controller
Fazlollahtabar & Saidi-Mehrabad, 2013	2013	Review	Not mentioned	Not mentioned	Not mentioned	Not mentioned	Not mentioned	Not mentioned	Optimisation through queueing network	Offline/Static batching but how?	Not mentioned	Not mentioned	Reviews exact approaches, meta-heuristics, and AI solutions	Reviews exact approaches, meta-heuristics, and AI solutions	Reviews exact approaches, meta-heuristics, and AI solutions
Ferrara et al., 2014	2014	Empirical	Laser-Guided Mobile Robots and Pallet Shuttles Mixed System	Put-away, Picking	Not mentioned	Not mentioned	Not mentioned	Studies double-warehouse	Not mentioned	Not mentioned	Not mentioned	Not mentioned	Reinforcement Learning enhanced by state aggregation	Not mentioned	Not mentioned
Camady et al., 2014	2014	Empirical	Not mentioned	Not mentioned	Not mentioned	Not mentioned	Not mentioned	Not mentioned	Not mentioned	Not mentioned	Not mentioned	Not mentioned	Not mentioned	Not mentioned	Zones to efficiently control the area
Ma et al., 2014	2014	Empirical	Not mentioned	Not mentioned	Not mentioned	AMR	Mixed	Not mentioned	Compares different fleet sizes and statistically proves the validity	Not mentioned	Not mentioned	Not mentioned	Not mentioned	D* with model predictive control to be assigned to the zones, A* in the zones	PSO variants (Con-Per-PSO and SA-PSO)
Digani et al., 2015	2015	Empirical	Not mentioned	Not mentioned	Not mentioned	Not mentioned	Not mentioned	Not mentioned	Not mentioned	Not mentioned	Not mentioned	Not mentioned	Not mentioned	D* with model predictive control to be assigned to the zones, A* in the zones	Negotiation (according to priority and zone change requests) with resource allocation strategy. Zoning with Voronoi decomposition

Study	Year	Type	Robot type	Application	Evaluation criteria	Performance metrics	Autonomy	Control architecture				Offline/Static, batching but how?	Compares different fleet sizes	Migration of task in case of failure		Genetic algorithm	Q-learning	Learning reward and prioritisation
Dou et al., 2015	2015	Empirical	Barcode-Guided Mobile Robots	Picking	Not mentioned	Robot utilisation / idleness, robot travel time	AMR	Mixed. Task Alloc. Centralised, Path Planning Distributed	Not mentioned	Not mentioned	Not mentioned	Offline/Static, batching but how?	Not mentioned	Not mentioned	Not mentioned	Not mentioned	Not mentioned	Not mentioned
Huang et al., 2015	2015	Review	Review barcode-guided and mobile picking robots	Not mentioned	Cost, material handling flexibility, scalability	Not mentioned	Not mentioned	Not mentioned	Not mentioned	Not mentioned	Not mentioned	Not mentioned	Not mentioned	Not mentioned	Not mentioned	Not mentioned	Not mentioned	Not mentioned
Kimura et al., 2015	2015	Empirical	Pick & Transport Robots	Picking	Not mentioned	Not mentioned	AMR	Not mentioned	Not mentioned	Not mentioned	Not mentioned	Not mentioned	Not mentioned	Not mentioned	Not mentioned	Not mentioned	Not mentioned	Not mentioned
Krnjak et al., 2015	2015	Empirical	Freeway Mobile Robots	Not mentioned	Not mentioned	Robot travel distance, number of deadlocks / conflicts / routing failures	AMR	Distributed	Not mentioned	Not mentioned	Not mentioned	Not mentioned	Not mentioned	Not mentioned	Not mentioned	Not mentioned	A* with static lattice	Checks for collision before each step
Vivaldini et al., 2015	2015	Review	Autonomous Forklifts	Not mentioned	Not mentioned	Not mentioned	AMR	Not mentioned	Not mentioned	Not mentioned	Not mentioned	Not mentioned	Not mentioned	Not mentioned	Not mentioned	Reviews online / offline allocation	Reviews dynamic and static algorithms	Reviews dynamic and static algorithms
Bauters et al., 2016	2016	Empirical	Barcode-Guided Mobile Robots	Picking	Cost, service quality, material handling flexibility, infrastructure flexibility, and scalability	Robot utilisation / idleness, workstation utilisation, throughput	AMR	Centralised	Not mentioned	Not mentioned	Not mentioned	Offline/Static, batching but how?	Not mentioned	Not mentioned	Not mentioned	Not mentioned	Not mentioned	Not mentioned
Bogue, 2016	2016	Review	Review barcode-guided, human-collaborated, and mobile picking robots	Picking	Not mentioned	Not mentioned	Not mentioned	Mixed. Macro-management is Centralised, inside of sectors are Distributed	Not mentioned	Not mentioned	Not mentioned	Not mentioned	Not mentioned	Not mentioned	Not mentioned	Not mentioned	Not mentioned	Not mentioned
Digani et al., 2016	2016	Empirical	Freeway Mobile Robots	Not mentioned	Not mentioned	Computation / negotiation time	AMR	Distributed	Not mentioned	Not mentioned	Not mentioned	Not mentioned	Compares different fleet sizes	Not mentioned	Not mentioned	One out of N algorithm	A*	Prioritisation. Zones to efficiently control the area
Draganjac et al., 2016	2016	Empirical	Autonomous Forklifts	Shipping	Not mentioned	Not mentioned	AMR	Distributed	Not mentioned	Not mentioned	Not mentioned	Not mentioned	Not mentioned	Migration of task in case of failure	Not mentioned	Not mentioned	State lattice construction and A*	Prioritisation and private zone – waiting or re-route

Reference	Year	Type	Technology	Process	(—)	Metrics	Robot type	Architecture	(—)	Human/Change	SKU effect	Dynamics	Fleet sizing	(—)	(—)	Task allocation	Path planning	Conflict / collision
Jin et al., 2016	2016	Empirical	Barcode-Guided Mobile Robots	Picking, Replenishment	Not mentioned	Not mentioned	AMR	Centralised	Not mentioned	Not mentioned	Not mentioned	Online/Dynamic	Not mentioned	Not mentioned	Not mentioned	Shortest distance + shortest task queue length of workstation	Not mentioned	Cost incurred if collision occurs
Ma & Koenig, 2016	2016	Empirical	Barcode-Guided Mobile Robots	Picking	Not mentioned	Not mentioned	AMR	Not mentioned	Not mentioned	Not mentioned	Not mentioned	Not mentioned	Compares different fleet sizes	Not mentioned	Not mentioned	Conflict-based min-cost-flow algorithm	Conflict-based min-cost-flow algorithm	Conflict-based min-cost-flow algorithm
Moeller et al., 2016	2016	Empirical	Not mentioned	Picking	Not mentioned	Robot travel distance	Not mentioned	Not mentioned	Not mentioned	Not mentioned	Not mentioned	Not mentioned	Not mentioned	Not mentioned	Not mentioned	Not mentioned	Not mentioned	Not mentioned
Santos et al., 2016	2016	Empirical	Freeway Mobile Robots	Not mentioned	Not mentioned	Robot travel distance, Avg task completion time, task density	AMR	Defines coordination types	Not mentioned	Change Management	Not mentioned	Not mentioned	Not mentioned	Not mentioned	Not mentioned	Not mentioned	Time Enhanced A*	Time Enhanced A*
Vivaldini et al., 2016	2016	Empirical	Autonomous Forklifts	Shipping	Not mentioned	Emission metrics, utilisation of robots, resource efficiency metrics	Mention electric and diesel robots next to battery robots	Centralised	Not mentioned	Not mentioned	Not mentioned	Not mentioned	Optimum number of robots through a mathematical model	Not mentioned	Not mentioned	Both tabu search with nearest neighbour and shortest job first	Enhanced Dijkstra's including turning times	A*
Bechtsis et al., 2017	2017	Review	Not mentioned	Not mentioned	Not mentioned	Not mentioned	AMR	Not mentioned	Not mentioned	Review human-robot interaction management t sub-sections	Studies the effect of SKU diversity on shelves	Not mentioned	Not mentioned	Not mentioned	Not mentioned	Mention online allocation strategies	Not mentioned	Not mentioned
Boysen et al., 2017	2017	Empirical	Barcode-Guided Mobile Robots	Picking	Not mentioned	Not mentioned	AMR	Not mentioned	Not mentioned	Not mentioned	Not mentioned	Offline/Static, batching with simulated annealing	Through SKU diversity in shelfs	Not mentioned	Not mentioned	Simulated annealing	Monte carlo tree search with iterative greedy heuristic	Not mentioned
Claes et al., 2017	2017	Empirical	Not mentioned	Not mentioned	Not mentioned	Avg task completion time, negotiation time	Not mentioned	Distributed	Not mentioned	Not mentioned	Not mentioned	Not mentioned	Not mentioned	Not mentioned	Not mentioned	Monte carlo tree search with iterative greedy heuristic	Not mentioned	Not mentioned
Dewangan et al., 2017	2017	Review	Not mentioned	Not mentioned	Not mentioned	Not mentioned	Not mentioned	Not mentioned	Not mentioned	Not mentioned	Not mentioned	Not mentioned	Not mentioned	Not mentioned	Not mentioned	Not mentioned	Reviews prioritised path planning algorithms	Defines conflict types
Fartielli et al., 2017	2017	Empirical	Barcode-Guided Mobile Robots	Picking	Not mentioned	Avg task completion time, negotiation time	AMR	Not mentioned	Not mentioned	Not mentioned	Not mentioned	Not mentioned	Compares different fleet sizes	Not mentioned	Not mentioned	Greedy local heuristic solving distributed constraint optimisation problem	Shortest path algorithms such as Dijkstra's	Prioritisation

Author	Year	Type	Technology															
Kamwichd a & Kazama, 2017	2017	Empirical	Barcode-Guided Mobile Robots	Picking	Not mentioned	Not mentioned	AMR	Not mentioned	Not mentioned	Not mentioned	Not mentioned	Not mentioned	Not mentioned	Not mentioned	Not mentioned	Not mentioned	Deep reinforcement learning	Not mentioned
Lambathali et al., 2017	2017	Empirical	Barcode-Guided Mobile Robots	Picking	Not mentioned	Throughput (time-based, order line-based), robot utilisation / idleness.	AMR	Not mentioned	Shelf and workstation layout studies through queueing theory	Not mentioned	ABC storage policy vs random storage policy	Online/Dynamic	Performance estimations through the number of mobile robots	Not mentioned	Not mentioned	Not mentioned	Dijkstra's	Zones to efficiently control the area.
Li et al., 2017	2017	Empirical	Barcode-Guided Mobile Robots	Picking	Not mentioned	Not mentioned	AMR	Distributed	Not mentioned	Not mentioned	Not mentioned	Offline/Static, batching through three-stage hybrid heuristic algorithm	Not mentioned	Not mentioned	Not mentioned	Mixed heuristic algorithm	Not mentioned	Not mentioned
Ma et al., 2017	2017	Empirical	Barcode-Guided Mobile Robots	Not mentioned	Not mentioned	Throughput (time-based, order line-based)	AMR	Not mentioned	Not mentioned	Not mentioned	Not mentioned	Online/Dynamic	Not mentioned	Not mentioned	Not mentioned	Once assigned, tasks could be swapped	Token passing with task swaps	Token passing with task swaps
Pagani et al., 2017	2017	Empirical	Freeway Mobile Robots	Not mentioned	Not mentioned	Not mentioned	AMR	Centralised	Not mentioned	Not mentioned	Not mentioned	Not mentioned	Not mentioned	Not mentioned	Not mentioned	Neural networks trained with genetic algorithm	Not mentioned	Not mentioned
R. Yan et al., 2017	2017	Empirical	Laser-Guided Mobile Robots	Not mentioned	Not mentioned	Not mentioned	Not mentioned	Not mentioned	Not mentioned	Not mentioned	Not mentioned	Online/Dynamic	Not mentioned	Identifies mission reliability of mobile robots via fault tree analysis and petri net approach	Not mentioned	Not mentioned	Not mentioned	Not mentioned
Sabattini et al., 2017	2017	Empirical	Autonomous Forklifts	Not mentioned	Not mentioned	Robot travel distance, travel time, avg task completion time, task delay / response time, number of deadlocks / conflicts / routing failures	AMR	Centralised	Not mentioned	Not mentioned	Not mentioned	Not mentioned	Compares different fleet sizes	Not mentioned	Not mentioned	Integer linear programming	Shortest path algorithms such as Dijkstra's	Conflict graph and linear programming
Yan et al., 2017	2017	Empirical	Barcode-Guided Mobile Robots	Picking	Not mentioned	Not mentioned	AMR	Centralised	Not mentioned	Not mentioned	Not mentioned	Not mentioned	Compares different fleet sizes	Not mentioned	Not mentioned	Not mentioned	Dijkstra's	
Yuan & Gong, 2017	2017	Empirical	Barcode-Guided Mobile Robots	Picking	Not mentioned	Throughput (time-based, order line-based), robot utilisation / idleness,	AMR	Centralised	Not mentioned	Not mentioned	Not mentioned	Not mentioned	Optimum number of robots through a mathematical model, dedicated	Not mentioned	Not mentioned	Not mentioned	Not mentioned	Not mentioned

Study	Year	Method	Technology	Application	Metrics	Robot Type	Task Allocation / Path Planning	Online/Dynamic	Fleet comparison	Charging	Assignment	Path algorithm	Other
Zou et al., 2017	2017	Empirical	Barcode-Guided Mobile Robots	Picking	workstation utilisation, robot, human ratio	AMR	Not mentioned	Online/Dynamic	vs pooled robots	Not mentioned	Near optimal assignment rule	Shortest path algorithms such as Dijkstra's	Uni-directional aisles
Abbas et al., 2018	2018	Empirical	AS/RS, Conveyor, Mobile Robots	Sorting	Throughput (time-based, order line-based), workstation utilisation	AGV	Not mentioned	Not mentioned	Compares different fleet sizes	Not mentioned	Not mentioned	Shortest path algorithms such as Dijkstra's	Not mentioned
Bechtsis et al., 2018	2018	Empirical	Autonomous Forklifts	Not mentioned	Emission metrics, utilisation /idleness of robots, resource efficiency metrics	AMR	Not mentioned	Not mentioned	Not mentioned	Not mentioned	Not mentioned	A*	Not mentioned
Chen et al., 2018	2018	Empirical	Freeway Mobile Robots	Sorting	Not mentioned	AMR	Mixed. Task Alloc. Centralised, Path Planning Distributed	Not mentioned	Not mentioned	Not mentioned	Not mentioned	Improved Artificial Potential Function	Improved Artificial Potential Function
De Koster, 2018	2018	Review	Reviews barcode-guided and human-collaborative/d mobile robots	Not mentioned	Not mentioned	Not mentioned	Not mentioned	Not mentioned	Not mentioned	Not mentioned	Not mentioned	Not mentioned	Not mentioned
Fan et al., 2018	2018	Empirical	Barcode Guided Mobile Robots	Sorting	Throughput (time-based, order line-based), number of deadlocks/ conflicts / routing failures, number of active robots	AGV	Centralised	Not mentioned	Minimises number of mobile robots required	Not mentioned	Not mentioned	Link weight increment with time window	Forward time window searching, waiting
Ghasemi & Chowdhur y, 2018	2018	Empirical	Barcode Guided Mobile Robots	Picking	Robot utilisation / idleness, computatio n / negotiation time	AMR	Distributed	Not mentioned	Not mentioned	Minimises the charging locations in the operational area through online connected dominating	Bipartite graph matching with fuzzy task clustering	Not mentioned	Not mentioned
Hamann et al., 2018	2018	Empirical	Barcode-Guided Mobile Robots	Picking	Not mentioned	AMR	Not mentioned	Not mentioned	Not mentioned	Not mentioned	Not mentioned	Not mentioned	Not mentioned

Reference	Year	Approach	Robot type	Process	Cost, material handling flexibility	Hit rate / number of deadlocks / conflicts / routing failures	AMR / AGV	Centralised / Distributed	Warehouse layouts	Ergonomics / human–robot risk	Storage	Online / Dynamic	Fleet sizes	Task migration / failure	Battery	Algorithm 1	Algorithm 2	Algorithm 3
Haasoon et al., 2018	2018	Empirical	Barcode-Guided Mobile Robots	Picking	Cost, material handling flexibility	Hit rate, number of deadlocks / conflicts / routing failures	AMR	Not mentioned	Not mentioned	Mention advantages of ergonomics	Not mentioned	Not mentioned	Not mentioned	Not mentioned	Not mentioned	set algorithm	Not mentioned	Not mentioned
He et al., 2018	2018	Empirical	Not mentioned	Not mentioned	Not mentioned	Not mentioned	Not mentioned	Not mentioned	Not mentioned	Not mentioned	Storage according to product service levels	Online / Dynamic	Impact of different fleet sizes	Not mentioned	Not mentioned	Deadline minimisation through Differentiated Probabilistic Queueing solved by simulated annealing and order swapping	Not mentioned	Not mentioned
Inam & Raizer, 2018	2018	Empirical	Human-collaborated mobile robots	Not mentioned	Not mentioned	Not mentioned	Not mentioned	Distributed	Vertical, horizontal, and fishbone layouts studied	Risk assessment in human-robot collaboration scenarios	Not mentioned	Not mentioned	Not mentioned	Not mentioned	Not mentioned	Not mentioned	Not mentioned	Not mentioned
Kattepur et al., 2018	2018	Empirical	Barcode-Guided Mobile Robots	Picking	Not mentioned	Not mentioned	AMR	Not mentioned	Not mentioned	Not mentioned	Not mentioned	Not mentioned	Not mentioned	Not mentioned	Not mentioned	Contract net protocol (negotiation)	A^*	Not mentioned
Kumar & Kumar, 2018	2018	Empirical	Barcode-Guided Mobile Robots	Picking	Not mentioned	Robot travel time	AMR	Not mentioned	Not mentioned	Not mentioned	Not mentioned	Not mentioned	Not mentioned	Migration of task in case of failure	Dual-decomposition to minimise battery consumption	Genetic algorithm	Oppositional-based learning + invasive weed optimisation	Not mentioned
Liu, 2018	2018	Empirical	AS/RS, Conveyor, Mobile Robots	Put-away, Picking	Not mentioned	Robot travel distance, travel time	AGV	Centralised	Not mentioned	Not mentioned	Not mentioned	Not mentioned	Not mentioned	Not mentioned	Not mentioned	Genetic algorithm	Oppositional-based learning + invasive weed optimisation	Voronoi diagram
Panda et al., 2018	2018	Empirical	Freeway Mobile Robots	Not mentioned	Not mentioned	Not mentioned	AMR	Not mentioned	Not mentioned	Not mentioned	Not mentioned	Not mentioned	Not mentioned	Not mentioned	Not mentioned	Not mentioned	Shortest path algorithms such as Dijkstra's	Local method based on velocity obstacle
Piccinelli et al., 2018	2018	Empirical	Self-Picking Mobile Robots	Picking	Not mentioned	Robot travel distance, robot utilisation / idleness, robot travel	AMR	Centralised	Comparisons under eight warehouse layouts	Not mentioned	Not mentioned	Not mentioned	Compares different fleet sizes	Not mentioned	Not mentioned	Shortest distance	Shortest path algorithms such as Dijkstra's	Re-route if the next grid is occupied
Qi et al., 2018	2018	Empirical	Barcode-Guided Mobile Robots	Not mentioned	Not mentioned	Not mentioned	AGV	Centralised	Not mentioned	Not mentioned	Not mentioned	Not mentioned	Not mentioned	Not mentioned	Not mentioned	Not mentioned	Shortest path algorithms such as Dijkstra's	First come first served, checks each time before deadlock, re-route. Zones to

Study	Year	Method	Robot Type	Application	Performance Metric	AMR	Control Architecture	Experimental Setup	Comparison	Order Handling	Fleet Size	Maintenance	Battery	Task Allocation / Algorithm	Path Planning / Routing	Congestion / Objective
R. Tai et al., 2018	2018	Empirical	Not mentioned	Not mentioned	Throughput (time-based, order line-based)	Not mentioned	Centralised	Not mentioned	Not mentioned	Online/Dynamic	Compares different fleet sizes	Not mentioned	Not mentioned	Shortest distance	k-shortest path planning	Time window with delays / efficiently control the area
R. Yan et al., 2018	2018	Empirical	Not mentioned	Not mentioned	Not mentioned	Not mentioned	Not mentioned	Not mentioned	Not mentioned	Not mentioned	Not mentioned	Preventive and corrective maintenance through petri nets and genetic algorithm	Not mentioned	Not mentioned	Not mentioned	Not mentioned
Sarkar et al., 2018	2018	Empirical	Self-Picking Mobile Robots	Picking	Computation / negotiation time	AMR	Not mentioned	Not mentioned	Optimisation through Nearest-neighbor based clustering and routing	Offline/Static, waves of orders	Not mentioned	Not mentioned	Not mentioned	Nearest-neighbor based clustering and routing	Nearest-neighbor based clustering and routing	Not mentioned
Senawai et al., 2018	2018	Empirical	Self-Picking Mobile Robots	Picking	Avg task completion time	AMR	Distributed	Not mentioned	Not mentioned	Online/Dynamic	Not mentioned	Not mentioned	Charges when battery is nearly 0%	A mechanism to sequentially execute interdependent tasks	Not mentioned	Not mentioned
Singhal et al., 2018	2018	Empirical	Not mentioned	Not mentioned	Avg task completion time	AMR	Mixed	Not mentioned	Compares different fleet sizes	Not mentioned	Not mentioned	Not mentioned	Not mentioned	Auction	Not mentioned	Not mentioned
Tsang et al., 2018	2018	Empirical	Not mentioned	Not mentioned	Not mentioned	AMR	Mixed. Task Alloc. Centralised, Path Planning Distributed	Not mentioned	Not mentioned	Not mentioned	Not mentioned	Not mentioned	Not mentioned	Genetic algorithm with learning heuristic	Recursive excitation/excitation artificial potential field Tabu temporal difference learning consisting of adaptive action selection rule and tabu action elimination strategy	Recursive excitation/excitation artificial potential field Tabu temporal difference learning consisting of adaptive action selection rule and tabu action elimination strategy
Wei & Ni, 2018	2018	Empirical	Guided Mobile Robots	Not mentioned	Robot travel distance	AMR	Distributed	Two aisles and aisle lengths with different number of workstations	Compares five storage assignment policies	Not mentioned	Not mentioned	Not mentioned	Not mentioned	Not mentioned	Adaptive large neighbourhood search	Congestion tracker and re-route
Weidinger et al., 2018	2018	Empirical	Barcode-Guided Mobile Robots	Picking	Robot travel distance, computation / negotiation time	AMR	Not mentioned	Not mentioned	Optimum number of robots through a mathematical model	Not mentioned	Not mentioned	Not mentioned	Not mentioned	Adaptive large neighbourhood search	Adaptive large neighbourhood search	Not mentioned

Study	Year	Type	Guidance / Navigation	Picking	Performance criteria	Performance metrics	Robot type	Control	Charging area	Human activity	Order type	Fleet sizing	Failure / Re-planning	Charging / Queueing	Allocation method	Routing	Preventive / Layout
Wen et al., 2018	2018	Review	Guided Mobile Robots	Not mentioned	Cost, material handling flexibility	Not mentioned	Not mentioned	Centralised	Not mentioned	Not mentioned	Not mentioned	Not mentioned	Not mentioned	Not mentioned	Not mentioned	Swarm intelligence	Not mentioned
Wiur et al., 2018	2018	Review	Categorise and review systems based on their navigation	Not mentioned	Cost, material handling flexibility, interruption behaviour (resilience)	Robot utilisation / idleness, workstation utilisation, throughput, avg task completion time, task delay / response time, number of deadlocks / conflicts / routing failures	Not mentioned	Not mentioned	Not mentioned	Not mentioned	Not mentioned	Not mentioned	Review the effect of failure on systems	Not mentioned	Not mentioned	Not mentioned	Review preventive and corrective ability of systems
Xue & Dong, 2018	2018	Empirical	Barcode-Guided Mobile Robots	Picking	Not mentioned	Avg task completion time	AMR	Not mentioned	Not mentioned	Not mentioned	Offline/Static, waves of orders	Not mentioned	Not mentioned	Not mentioned	Improved ant colony optimisation	Not mentioned	Not mentioned
Y Liu et al., 2018	2018	Empirical	Freeway Mobile Robots	Not mentioned	Not mentioned	Not mentioned	Not mentioned	Decentralised (Mixed)	Not mentioned	Not mentioned	Not mentioned	Not mentioned	Re-calculate task and path planning	Not mentioned	Auction	Incidental delivery	Not mentioned
Z. Liu et al., 2018	2018	Review	Not mentioned	Not mentioned	Not mentioned	Not mentioned	Not mentioned	Not mentioned	Not mentioned	Review vision, sensor, and radio frequency-based human activity recognition	Not mentioned	Not mentioned	Not mentioned	Not mentioned	Mention online allocation strategies	Not mentioned	Selecting other route, waiting before starting, route modification, re-dispatching
Z. Zhang et al., 2018	2018	Empirical	Barcode-Guided Mobile Robots	Picking	Not mentioned	Avg task completion time	AMR	Centralised	Not mentioned	Not mentioned	Not mentioned	Not mentioned	Not mentioned	Not mentioned	Prioritisation	Improved Dijkstra's	Not mentioned
Zavoudakis et al., 2018	2018	Empirical	Laser-Guided and Autonomous Forklifts	Not mentioned	Not mentioned	Not mentioned	Define seven criteria and apply R-Rov method to select among alternatives	Not mentioned	Considers placing charging locations in operational area	Not mentioned	Not mentioned	Not mentioned	Not mentioned	Charge when energy level is below 20%	Not mentioned	Not mentioned	Uni-directional aisles
Zou et al., 2018	2018	Empirical	Barcode-Guided Mobile Robots	Picking	Cost, material handling flexibility, scalability	Throughput (time-based, order line-based), workstation utilisation	AMR	Not mentioned	Not mentioned	Not mentioned	Not mentioned	Finds required number of robots for instances	Not mentioned	Semi-open queueing network considering throughput performance	Not mentioned	Shortest rectangular path	Not mentioned

Author, Year	Type	Scope	Task	Criteria	Metrics	Robot	Control	Layout		Storage		Static/Dynamic	Fleet			Task allocation	Path planning	Zone management
Azadeh et al., 2019	Review	Review barcode-guided, human-collaborated mobile robots	Picking, Replenishment	Not mentioned	Throughput (time-based, order lines-based), robot utilisation / idleness, task delay / response time	Not mentioned	Not mentioned	Not mentioned	Not mentioned	Mention strategies	Not mentioned	Not mentioned	Not mentioned	Not mentioned	Not mentioned	Mention strategies	Not mentioned	Not mentioned
Bormann et al., 2019	Empirical	Self-Picking Mobile Robots	Picking	Not mentioned	Not mentioned	AMR	Not mentioned	Not mentioned	Not mentioned	Not mentioned	Not mentioned	Not mentioned	Not mentioned	Not mentioned	Not mentioned	Not mentioned	Shortest path algorithms such as Dijkstra's	Not mentioned
Boysen et al., 2019	Review	Review barcode-guided, human-collaborated mobile robots	Picking, Replenishment, Sorting	Define criteria for e-commerce setting	Not mentioned	Not mentioned	Not mentioned	Not mentioned	Not mentioned	Review different policies	Not mentioned	Review static and dynamic strategies	Not mentioned	Not mentioned	Not mentioned	Not mentioned	Not mentioned	Prioritisation s
D'Tadello & Khan, 2019	Empirical	Guided Mobile Robots	Not mentioned	Not mentioned	Robot travel distance, Avg task completion time	AMR	Mixed	Compares different number of aisles and lengths	Not mentioned	Not mentioned	Not mentioned	Online/Dynamic	Compares different fleet sizes	Not mentioned	Not mentioned	Auction	Safe Interval Path Planning	Prioritisation n. Zones to efficiently control the area
Digani et al., 2019	Empirical	Self-Picking Mobile Robots	Picking	Not mentioned	Robot travel time, avg task completion time	AMR	Centralised	Not mentioned	Not mentioned	Not mentioned	Not mentioned	Not mentioned	Not mentioned	Not mentioned	Not mentioned	Waiting time minimisation through negotiation	Quadratic optimisation with A*	Not mentioned
Feng et al., 2019	Empirical	Barcode-Guided Mobile Robots	Picking	Not mentioned	Robot travel distance	AMR	Not mentioned	Compares Flying-V and Traditional layouts with different number of aisles and aisle lengths	Not mentioned	Random storage assignment policy	Not mentioned	Not mentioned	Not mentioned	Not mentioned	Not mentioned	0-1 integer programming	Shortest path algorithms such as Dijkstra's	Not mentioned
Fazeler et al., 2019	Empirical	Rail Using Mobile Robots	Picking	Not mentioned	Not mentioned	AGV	Not mentioned	Not mentioned	Not mentioned	SKU storage optimisation through Mixed Integer Programming via adjusted Simulated Annealing	Not mentioned	Not mentioned	Not mentioned	Not mentioned	Not mentioned	Priority-rule-based approach, order swapping	Not mentioned	Zones to efficiently control the area
H. Wang et al., 2019	Empirical	Human-collaborated mobile robots	Picking	Not mentioned	Throughput (time-based, order lines-based), robot utilisation / idleness, robot travel time	AMR	Not mentioned	Not mentioned	Not mentioned	Not mentioned	Not mentioned	Offline/Static ic, batching but how?	Compares different fleet sizes	Not mentioned	Not mentioned	Genetic algorithm	A*	Reservation of routes through time windows

Reference	Year	Type	Robot type	Application	Economic metrics	Performance KPIs	Robot class	Warehouse layout	Control architecture	Storage policy	Scheduling	Fleet sizing	Failure handling	Charging	Path planning	Path planning / optimisation	Intersection management
Haiming et al., 2019	2019	Empirical	Barcode-Guided Mobile Robots	Picking	Not mentioned	Robot travel distance, travel time, number of deadlocks / conflicts / failures	AMR	Not mentioned	Not mentioned	Not mentioned	Online/Dynamic	Compares different fleet sizes	Not mentioned	Not mentioned	Shortest response time	Dijkstra's with dynamic path costs update method	Treat intersections at static standpoints and manage them with rules
Hirayama & Nagao, 2019	2019	Empirical	Linear Route Mobile Robots	Not mentioned	Not mentioned	Not mentioned	Not mentioned	Not mentioned	Not mentioned	Not mentioned	Not mentioned	Compares different fleet sizes	Not mentioned	Not mentioned	Not mentioned	Not mentioned	Not mentioned
Hornakova et al., 2019	2019	Empirical	Laser-Guided Autonomous Forklifts	Not mentioned	Mobile robot cost, infrastructure flexibility	Not mentioned	Define AGV, AMR, and seven criteria to select among alternatives. Use AHP method	Not mentioned	Not mentioned	Not mentioned	Not mentioned	Not mentioned	Not mentioned	Not mentioned	Not mentioned	Not mentioned	Not mentioned
K. Wang et al., 2019	2019	Empirical	Barcode-Guided Mobile Robots	Picking	Not mentioned	Not mentioned	AMR	Six scenarios with different number of aisles and layers. Single vs double-deep shelf layouts	Not mentioned	Not mentioned	Not mentioned	Not mentioned	Not mentioned	Not mentioned	Not mentioned	Not mentioned	Not mentioned
Lee & Murray, 2019	2019	Empirical	Pick & Transport Robots	Picking	Not mentioned	Robot travel distance, computation / negotiation time	AMR	Compares two types of layouts with different number of workstations and storages	Centralised	Uniform storage policy	Offline/Static: batching but how?	Compares different fleet sizes	Not mentioned	Charges when battery is nearly 0%	Linear programming	Linear programming	Time window to calculate, waiting, before starting, re-route to another optimal route and go-away to prevent
Lee et al., 2019	2019	Empirical	Barcode-Guided Mobile Robots	Picking, Replenishment	Not mentioned	Throughput (time-based, order line-based)	AMR	Not mentioned	Not mentioned	Not mentioned	Not mentioned	Not mentioned	Not mentioned	Charge when energy level is below 40%	Not mentioned	Improved A* using manhattan distance and reserving all possible routes and including turning times	Reservations of routes through time windows
Lienert et al., 2019	2019	Empirical	Barcode-Guided Mobile Robots	Picking	Not mentioned	Robot utilisation, idleness, robot travel time	AMR	Not mentioned	Not mentioned	Not mentioned	Not mentioned	Compares different fleet sizes	Re-route if a robot fails	Not mentioned	Not mentioned	A*	Not mentioned
Liagkou-Albert et al., 2019	2019	Empirical	Not mentioned	Not mentioned	Mobile robot cost, implementation time, scalability	Not mentioned	AMR	Not mentioned	Not mentioned	Not mentioned	Not mentioned	Not mentioned	Not mentioned	Not mentioned	Not mentioned	Not mentioned	Waiting, re-routing through time windows-based

Study	Year	Type																conflict detection
Ly, 2019	2019	Empirical	Not mentioned	Not mentioned	Not mentioned	Not mentioned	Not mentioned	Not mentioned	Not mentioned	Not mentioned	ABC zoning storage policy and closest open location storage within the zones	Not mentioned	Not mentioned	Not mentioned	Not mentioned	Shortest distance	k-shortest path planning	Windowed Hierarchica Cooperative A*
Mernescher mann et al., 2019	2019	Empirical	Barcode-Guided Mobile Robots	Picking, Replenishment	Not mentioned	Robot travel distance, robot utilisation / idleness, workstation utilisation	AMR	Not mentioned	Not mentioned	Not mentioned	Not mentioned	Not mentioned	Not mentioned	Not mentioned	Not mentioned	Not mentioned	Windowed Hierarchica Cooperative A*	Not mentioned
Otto & Ishiguni, 2019	2019	Empirical	Human-collaborated mobile robots	Picking	Not mentioned	Not mentioned	AMR	Not mentioned	Not mentioned	Not mentioned	Study five different rules for storage assignment	Offline/Static, batches replenishment orders via random or shelf-based batching	Compares different fleet sizes per workstation	Not mentioned	Not mentioned	Mixed integer linear programming solved by an algorithm	Shortest path algorithms such as Dijkstra's	Not mentioned
P Li a., 2019	2019	Empirical	Autonomous Forklifts	Put-away, Picking	Not mentioned	Robot travel distance, robot travel time, task delay / response time	AMR	Centralised	Not mentioned	Not mentioned	Not mentioned	Not mentioned	Mixed integer linear programming	Not mentioned	Not mentioned	Deep Q-Network (reinforcement learning)	A*	Deep Q-Network (reinforcement learning) rewards for staying away from the traffic
Papcun et al., 2019	2019	Empirical	Freeway Mobile Robots	Picking	Not mentioned	Not mentioned	AMR	Not mentioned	Not mentioned	Human workers follow mobile robot path through augmented reality glasses Human intention estimation through hidden markov model and bayesian theory of mind approach	Not mentioned	Not mentioned	Not mentioned	Not mentioned	Not mentioned	Not mentioned	Shortest path algorithms such as Dijkstra's	Not mentioned
Petkovic et al., 2019	2019	Empirical	Barcode-Guided Mobile Robots	Picking	Not mentioned	Not mentioned	Not mentioned	Not mentioned	Not mentioned	Safety through velocity planner	Not mentioned	Not mentioned	Not mentioned	Not mentioned	Not mentioned	Not mentioned	Voronoi diagram based path planning	Not mentioned
Ratnert et al., 2019	2019	Empirical	Laser-Guided Mobile Robots	Not mentioned	Not mentioned	Not mentioned	Not mentioned	Not mentioned	Not mentioned	Not mentioned	Not mentioned	Not mentioned	Not mentioned	Not mentioned	Not mentioned	Not mentioned	Not mentioned	Estimates system performance with and without zones

Author	Year	Type	Robot type	Picking	Evaluation criteria	Performance metric	Robot class	Control	Layout	Storage policy		Online/Offline	Fleet	Task migration		Scheduling	Path planning	Routing / conflict
Rivas et al., 2019	2019	Empirical	Freeway Mobile Robots	Not mentioned	Not mentioned	Not mentioned	AMR	Decentralised (Mixed)	Not mentioned	Not mentioned	Not mentioned	Online/Dynamic	Compares different fleet sizes and finds the optimal one	Not mentioned	Not mentioned	Auction with Dijkstra's to calculate the distance	Shortest path algorithms such as Dijkstra's	Not mentioned
Roy et al., 2019	2019	Empirical	Barcode-Guided Mobile Robots	Picking	Cost, material handling flexibility, scalability	Throughput (time-based, order line-based)	AMR	Not mentioned	Not mentioned	Random open location shelf storage policy	Not mentioned	Not mentioned	Only compares dedicated vs pooled fleets in zones	Not mentioned	Not mentioned	First come first served with a multi-class closed queueing model	Not mentioned	Not mentioned
Sarkar & Agarwal, 2019	2019	Empirical	Self-Picking Mobile Robots	Picking	Not mentioned	Deadline misses	AMR	Centralised	Not mentioned	Not mentioned	Not mentioned	Offline/Static, waves of orders	Not mentioned	Not mentioned	Not mentioned	Minimum penalty scheduling	Not mentioned	Cost incurred if collision occurs
Sartoretti et al., 2019	2019	Empirical	Freeway Mobile Robots	Not mentioned	Not mentioned	Not mentioned	AMR	Mixed	Not mentioned	Not mentioned	Not mentioned	Not mentioned	Not mentioned	Migration of task in case of failure	Not mentioned	Not mentioned	Not mentioned	Defines conflict types
Stern, 2019	2019	Review	Not mentioned	Not mentioned	Not mentioned	Not mentioned	Not mentioned	Not mentioned	Compares two types of layouts with different number of workstations and storages	Not mentioned	Not mentioned	Not mentioned	Not mentioned	Not mentioned	Not mentioned	Not mentioned	Groups algorithms under four titles and reviews them	Time windows - Waiting before moving to the next grid if it is occupied and prioritisation. Insert Route algorithm with solution improvement at via late acceptance hill climbing
Tai et al., 2019	2019	Empirical	Barcode-Guided Mobile Robots	Not mentioned	Not mentioned	Throughput (time-based, order line-based), computation / negotiation time	Not mentioned	Centralised	Not mentioned	Not mentioned	Not mentioned	Online/Dynamic	Compares different fleet sizes	Not mentioned	Not mentioned	Shortest makespan	Reinforcement Learning combined with imitation learning (OD-recursive M*)	Insert Route algorithm with solution improvement at via late acceptance hill climbing. Waiting-re-routing with deadlock detection algorithm. Zones to efficiently control the area
Thamer et al., 2019	2019	Empirical	Freeway Mobile Robots	Not mentioned	Not mentioned	Not mentioned	AMR	Not mentioned	Not mentioned	Not mentioned	Not mentioned	Not mentioned	Not mentioned	Not mentioned	Not mentioned	Static last in first out	k-shortest path planning	Insert Route algorithm with solution improvement at via late acceptance hill climbing
Tsolakis et al., 2019	2019	Empirical	Autonomous Forklifts	Not mentioned	Not mentioned	Not mentioned	Not mentioned	Not mentioned	Not mentioned	Not mentioned	Not mentioned	Not mentioned	Not mentioned	Not mentioned	Not mentioned	Not mentioned	Not mentioned	Not mentioned

Reference	Year	Type																	
Wu & Ge, 2019	2019	Review	Barcode-Guided Mobile Robots	Picking	Not mentioned	Not mentioned	AMR	Not mentioned	Not mentioned	Not mentioned	Not mentioned	Not mentioned	Not mentioned	Not mentioned	Not mentioned	Not mentioned	Not mentioned	Not mentioned
X. Li et al., 2019	2019	Empirical	Not mentioned	Not mentioned	Not mentioned	Not mentioned	Not mentioned	Centralised	Not mentioned	Not mentioned	Not mentioned	Not mentioned	Not mentioned	Compares different fleet sizes	Not mentioned	Not mentioned	Shortest response time (idles vs closest to finish)	k-shortest path planning	Not mentioned
Xu et al., 2019	2019	Empirical	Not mentioned	Not mentioned	Not mentioned	Not mentioned	AMR	Mixed: Charging, Distributed, Task Alloc. Centralised	Not mentioned	Not mentioned	Not mentioned	Not mentioned	Not mentioned	Not mentioned	Not mentioned	Task sequencing with bipartite graph matching and time windows	Heuristic distributed task allocation	Not mentioned	Not mentioned
Y. Lin et al., 2019	2019	Empirical	Freeway Mobile Robots	Sorting	Not mentioned	Not mentioned	AMR	Not mentioned	Not mentioned	Not mentioned	Not mentioned	Not mentioned	Not mentioned	Minimises number of mobile robots required	Not mentioned	Creates a charging task according to power consumed by active time and speed	Multi-adaptive genetic algorithm	Not mentioned	Dynamic Weight Guidance
Y. Liu et al., 2019	2019	Empirical	Barcode-Guided Mobile Robots	Picking	Not mentioned	Not mentioned	AGV	Centralised	Not mentioned	Not mentioned	Not mentioned	Not mentioned	Not mentioned	Not mentioned	Re-calculate task and path planning	Not mentioned	Not mentioned	Improved Cooperative A*	Not mentioned
Yin & Xin, 2019	2019	Empirical	Freeway Mobile Robots	Not mentioned	Not mentioned	Not mentioned	Not mentioned	Not mentioned	Not mentioned	Not mentioned	Not mentioned	Not mentioned	Not mentioned	Not mentioned	Not mentioned	Not mentioned	Time-space network model optimisation	Time-space network model optimisation	Waiting before starting
Y. Zhang et al., 2019	2019	Empirical	Not mentioned	Not mentioned	Not mentioned	Not mentioned	Not mentioned	Not mentioned	Not mentioned	Not mentioned	Not mentioned	Not mentioned	Not mentioned	Not mentioned	Not mentioned	Not mentioned	Not mentioned	Improved A*	Not mentioned
J. Zhang et al., 2019	2019	Empirical	Barcode-Guided Mobile Robots	Picking	Not mentioned	Not mentioned	AMR	Not mentioned	Not mentioned	Not mentioned	Not mentioned	Not mentioned	Not mentioned	Not mentioned	Not mentioned	Recharge if below 50%	Genetic algorithm & Priority rule-based heuristic	Not mentioned	Not mentioned
Yoshitake et al., 2019	2019	Empirical	Barcode-Guided Mobile Robots	Picking	Not mentioned	Workstation utilisation	AMR	Mixed: Task Alloc. Centralised, Path Planning Distributed	Not mentioned	Not mentioned	Not mentioned	Not mentioned	Not mentioned	Not mentioned	Not mentioned	Not mentioned	Real-time holonic scheduling	Real-time holonic scheduling	Not mentioned
Yuan et al., 2019	2019	Empirical	Barcode-Guided Mobile Robots	Picking	Not mentioned	Robot travel distance	AMR	Not mentioned	Not mentioned	Not mentioned	Studies how many products on which shelf and shelf location via simulated annealing	Not mentioned	Not mentioned	Not mentioned	Not mentioned	Not mentioned	Not mentioned	Two-stage Mixed algorithm: Greedy algorithm and simulated annealing	Not mentioned

Study	Year	Research type	Robot type	Process		Performance metrics	Robot guidance	Control architecture							Task allocation	Path planning	Congestion management
Z. Liu et al.	2019	Empirical	Freeway Mobile Robots	Not mentioned	Not mentioned	Not mentioned	Not mentioned	Mixed	Not mentioned	Not mentioned	Not mentioned	Not mentioned	Not mentioned	Not mentioned	Auction	Incidental delivery	Zones to efficiently control the area
Zou et al.	2019	Empirical	Human-collaborated mobile robots	Picking	Not mentioned	Not mentioned	AMR	Not mentioned	Not mentioned	Not mentioned	Not mentioned	Not mentioned	Not mentioned	Not mentioned	Two stage heuristic algorithm	Two-stage heuristic algorithm refined by neighbourhood search heuristic	Prioritisation and re-routing
Dragasjac et al.	2020	Empirical	Autonomous Forklifts	Shipping	Not mentioned	Throughput (time-based, order line-based), robot utilisation / idleness, computation / negotiation time, number of deadlock / conflicts / routing failures	AMR	Distributed	Not mentioned	Not mentioned	Not mentioned	Not mentioned	Not mentioned	Not mentioned	One out of N algorithm	Shortest path algorithms such as Dijkstra's	Uni-directional aisles
Lamballais et al.	2020	Empirical	Barcode-Guided Mobile Robots	Picking	Not mentioned	Not mentioned	AMR	Not mentioned	Studies on proportion of picking and replenishment workstations	Optimises number of shelves per SKU and shelf replenishment level through queueing theory	Offline/Static batching with seed algorithm	Compares different fleet sizes	Not mentioned	Not mentioned	Not mentioned	Shortest path algorithms such as Dijkstra's	Not mentioned
Li et al.	2020	Empirical	Barcode-Guided Mobile Robots	Picking	Not mentioned	Not mentioned	AMR	Not mentioned	Not mentioned	Not mentioned	Not mentioned	Not mentioned	Not mentioned	Not mentioned	Particle swarm optimisation	Shortest path algorithms such as Dijkstra's	Gaussian mixture-based background segmentation algorithm for detection, re-route
Lee et al.	2020	Empirical	Not mentioned	Not mentioned	Not mentioned	Avg task completion time, computation / negotiation time	AMR	Decentralised (Mixed)	Not mentioned	Not mentioned	Not mentioned	Not mentioned	Not mentioned	Not mentioned	Auction-based winner determination on through Genetic algorithm	Genetic algorithm solved auction-based winner determination on	Not mentioned
Ng et al.	2020	Empirical	Freeway Mobile Robots	Picking	Not mentioned	Not mentioned	Not mentioned	Not mentioned	Not mentioned	Not mentioned	Not mentioned	Not mentioned	Not mentioned	Not mentioned	Not mentioned	A* (adaptive dynamic path finding)	Re-route if the aisle is occupied by another forklift or prioritisation
Pollten & Emde	2020	Empirical	Autonomous Forklifts	Picking, Put-away	Not mentioned	Not mentioned	AMR	Not mentioned	Optimised layout studies for narrow aisle warehouses	Not mentioned	Not mentioned	Number of forklifts are advised to be lower than number of aisles when	Not mentioned	Not mentioned	Large neighbourhood search and lateness minimisation	Not mentioned	Not mentioned

Wang et al., 2020	2020	Empirical	Barcode-Guided Mobile Robots	Picking	Not mentioned	Throughput (time-based, order line-based), robot travel distance	AMR	Not mentioned	Through a design framework with queueing, three different layouts	Not mentioned	Not mentioned	Online/Dynamic	aisles are narrow	Not mentioned	Not mentioned	Not mentioned	Not mentioned	Not mentioned
Witczak et al., 2020	2020	Empirical	Autonomous Forklifts	Shipping	Not mentioned	Not mentioned	AMR	Not mentioned	Not mentioned	Not mentioned	Not mentioned	Not mentioned	Not mentioned	Fault-tolerant control algorithm in case of failure	Not mentioned	Fault-tolerant control algorithm	Not mentioned	Not mentioned
Xing et al., 2020	2020	Empirical	Self-Picking Mobile Robots	Picking	Not mentioned	Not mentioned	AMR	Not mentioned	Not mentioned	Not mentioned	Not mentioned	Not mentioned	Not mentioned	Not mentioned	Not mentioned	Shortest distance	Tabu search and mission group exchange with neighbourhood search	Mission swapping and waiting

INDEX